AF841948

POLLUTION SCIENCE, TECHNOLOGY AND ABATEMENT

FORMALDEHYDE

CHEMISTRY, APPLICATIONS AND ROLE IN POLYMERIZATION

POLLUTION SCIENCE, TECHNOLOGY AND ABATEMENT

Additional books in this series can be found on Nova's website
under the Series tab.

Additional e-books in this series can be found on Nova's website
under the e-book tab.

CHEMISTRY RESEARCH AND APPLICATIONS

Additional books in this series can be found on Nova's website
under the Series tab.

Additional e-books in this series can be found on Nova's website
under the e-book tab.

POLLUTION SCIENCE, TECHNOLOGY AND ABATEMENT

FORMALDEHYDE

CHEMISTRY, APPLICATIONS AND ROLE IN POLYMERIZATION

CHAN BAO CHENG
AND
FENG HU LN
EDITORS

nova publishers
New York

Copyright © 2012 by Nova Science Publishers, Inc.

All rights reserved. No part of this book may be reproduced, stored in a retrieval system or transmitted in any form or by any means: electronic, electrostatic, magnetic, tape, mechanical photocopying, recording or otherwise without the written permission of the Publisher.

For permission to use material from this book please contact us:
Telephone 631-231-7269; Fax 631-231-8175
Web Site: http://www.novapublishers.com

NOTICE TO THE READER

The Publisher has taken reasonable care in the preparation of this book, but makes no expressed or implied warranty of any kind and assumes no responsibility for any errors or omissions. No liability is assumed for incidental or consequential damages in connection with or arising out of information contained in this book. The Publisher shall not be liable for any special, consequential, or exemplary damages resulting, in whole or in part, from the readers' use of, or reliance upon, this material. Any parts of this book based on government reports are so indicated and copyright is claimed for those parts to the extent applicable to compilations of such works.

Independent verification should be sought for any data, advice or recommendations contained in this book. In addition, no responsibility is assumed by the publisher for any injury and/or damage to persons or property arising from any methods, products, instructions, ideas or otherwise contained in this publication.

This publication is designed to provide accurate and authoritative information with regard to the subject matter covered herein. It is sold with the clear understanding that the Publisher is not engaged in rendering legal or any other professional services. If legal or any other expert assistance is required, the services of a competent person should be sought. FROM A DECLARATION OF PARTICIPANTS JOINTLY ADOPTED BY A COMMITTEE OF THE AMERICAN BAR ASSOCIATION AND A COMMITTEE OF PUBLISHERS.

Additional color graphics may be available in the e-book version of this book.

Library of Congress Cataloging-in-Publication Data

Formaldehyde : chemistry, applications, and role in polymerization / [edited by] Chan Bao Cheng and Feng Hu Lin.
pages cm
Includes bibliographical references and index.
ISBN 978-1-62257-214-4 (hardcover)
1. Formaldehyde. I. Cheng, Chan Bao, 1969- editor of compilation. II. Lin, Feng Hu, 1964- editor of compilation.
TP248.F6F67 2012
615.9'51--dc23

2012021148

Published by Nova Science Publishers, Inc. † New York

CONTENTS

PREFACE

Formaldehyde is a building block in the synthesis of many other compounds of specialized and industrial significance. It exhibits most of the chemical properties of other aldehydes but is more reactive. In this book, the authors discuss the chemistry, applications and role in polymerization of formaldehyde. Topics discussed include the properties of urea-formaldehyde resins for wood-based composites; electronic spectra of formaldehyde in aqueous solution; decontamination of indoor air pollutants of formaldehyde through catalytic oxidation over oxide supported noble metal nanocatalysts; indoor air monitoring using newly developed formaldehyde sensor elements and portable monitoring devices; unusual behavior during the electrochemical oxidation of formaldehyde; and an algebraic approach to estimate the PES of formaldehyde through the study of vibrational excitations.

Chapter 1 - This chapter reviews recent progresses on properties, chemical structure, thermal curing behavior, hydrolytic stability, morphology, microstructure, crystalline structure, and modifications of urea-formaldehyde (UF) resin as an adhesive for wood-based composite panels, particularly by focusing on the parameters related to formaldehyde emission (FE), such as synthesis reaction pH conditions, formaldehyde/urea (F/U) mole ratio, and resin modifications.

The reaction pH condition of UF resin synthesis showed that the amount of free formaldehyde strongly affected the reactivity of UF resin, and also indicated that the weak acid reaction condition provided a balance between increasing resin reactivity and improving adhesion strength of UF resins. Solid-state ^{13}C-NMR spectroscopy indicated that the molecular mobility of cured UF resin increased with decreasing the reaction pH used during its synthesis. The ^{13}C-NMR spectroscopy showed that UF resins with higher F/U mole ratios (i.e., 1.6 and 1.4) had two distinctive peaks, indicating the presence of dimethylene ether linkages and methylene glycols, which give a greater contribution to the FE than that of lower F/U mole ratio. However, these peaks were not detected at the UF resins with lower F/U mole ratios (i.e., 1.2 and 1.0). Lowering F/U mole ratio of UF resins as a way of abating FE consequently requires improving their reactivity. As the F/U mole ratio decreases, thermal curing behavior of these UF resins such as the gel time, onset and peak temperatures, and heat of reaction (ΔH) increased, while the activation energy (Ea) and rate constant (*k*) were decreased. The results also suggested that as the F/U mole ratio decreased, the FE of particleboard (PB) was greatly reduced at the expense of the reactivity of UF resin and slight deterioration of performance of PB prepared. Dynamic mechanical analysis (DMA) results partially explained the reason why UF resin adhesives with lower F/U mole ratio resulted in relatively poor adhesion performance.

Morphological investigation on UF resins illustrated that the spherical structures in cured UF resins were much more resistant to the hydrolytic degradation by the acid than amorphous region. Atomic force microscopy (AFM) images showed two distinctive regions, i.e., hard and soft phases in cured UF resins. The AFM study suggested that the soft phase was much more susceptible to the hydrolysis of cured UF resin than the hard phase. The soft phase of cured UF resins by ammonium chloride was much more easily hydrolyzed than those cured by ammonium sulfate, indicating that hardener types had a great impact on the hydrolytic degradation behavior of cured UF resins. For the first time, the presence of thin filament-like crystalline structures on the fracture surface of cured UF resin was reported. And X-ray diffraction (XRD) results showed that the crystalline regions of cured UF resins with lower F/U mole ratio contribute partially to the improved hydrolytic stability of the cured resin.

Chapter 2 - Formaldehyde is an important chemical feedstock for the production of phenoplast and aminoplast thermosetting resins, by reaction with other monomers (mostly urea, but also melamine, phenol and resorcinol). These adhesives are mainly used in the manufacture of wood-based panels: plywood, particleboard, hardboard, medium density fiberboard (MDF) and oriented strand board (OSB). These products have a wide range of applications, from non-structural to structural, outdoor or indoor, mostly in construction and furniture, but also in decoration and packaging. The WBP industry plays an important role in the global economy and contributes for forest sustainability and carbon sequestration. In 2009, FAO (Food and Agriculture Organization) reported that a total of 260 million m^3 WBPs were produced in the world (Europe 29.7%, Asia 43.9%, North America 18.3% and others 2.5%).

Being economically competitive and highly performing, a major drawback of formaldehyde-based resins, mostly urea-formaldehyde, is the formaldehyde emission during panel manufacturing and service life. There are two sources of emission: release of unreacted monomer, during or after panel production, and long-term resin degradation (hydrolysis). The formaldehyde content and chemical stability of the resin will therefore affect emission levels. In addition, external factors like temperature, humidity or air renewal rate will also play a role. It must be noted that wood itself contributes to formaldehyde emission, since it is a product of metabolism and decomposition processes. The actual emission level depends strongly on the type(s) of wood used in panel production.

Due to information considering formaldehyde as potentially carcinogenic to humans, the implementation of international regulations and requirements for emissions from WBPs has led to establishment of standard testing methods. Two main groups are considered: chamber methods (emulating indoor living environments, mentioned in ASTM, ISO and European standards), and small scale methods, also called derived tests, oriented to industrial quality control and development. This second group includes commonly used methods, mentioned in different international standards, like the so-called: perforator (actually a test of potential formaldehyde emission), flask, desiccator, and gas analysis methods. Correlation between results from different methods has been a matter of debate, not yet completely elucidated.

Based on different test methods, emission limit standards for WBPs have been issued by several governmental organizations in Europe, Japan and United States, allowing for product classification according to emission level. Additionally, limits drawn by major industrial consumers, like IKEA, have been a defining guideline for WBP producers.

In order to comply with increasingly stringent requirements, the industry has been developing strategies to minimize formaldehyde emissions from WBPs. Four major approaches can be found: 1) reduction of formaldehyde content in resin formulation, while attempting to maintain adhesive performance, 2) addition of formaldehyde scavengers to resin or wood particles, having the negative effect of consuming formaldehyde prior to resin cure, 3) implementation of surface treatments after board production, and 4) use of alternative adhesive systems with reduced or no emissions, with an impact on product cost and/or performance.

Chapter 3 - Two models are presented to estimate the electronic spectra for formaldehyde in condensed phase. Different from others' concerns, the key of our models is the establishment of proper energy expression to describe the Franck–Condon state with the aid of the constrained equilibrium method under the thermodynamics theory. In the first explicit solvent model, high-level quantum mechanics theory is employed to calculate formaldehyde and classical molecular dynamics method is adopted to simulate the individual solvent molecules. Mutual polarization between the two portions is adequately considered. The long–range electrostatic effect and short–range dispersion/repulsion effect in the solute–solvent system are introduced into the solute Hamiltonian as perturbation operators. In the second implicit solvent model, the solute formaldehyde with the point dipole approximation is located at the center of a spherical cavity surrounded by continuous dielectric medium and the cavity radius is determined by the molecular dynamics simulation. Both the two models well predict the solvatochromic shift of the singlet $n \rightarrow \pi^*$ transition for formaldehyde in aqueous solution. Different contributions to the total solvation shift are analyzed and the main component comes from the electrostatic plus polarization interactions. The microscopic solvent structure is quite disturbed by formaldehyde to form three solvation shells. There are mainly three or two dynamic hydrogen bonds formed between formaldehyde and water molecules. Since the hydrogen bonding effect is always deemed to be the dominant contribution to the solvatochromic shift of polar chromophores in aqueous solution, the lowest singlet $n \rightarrow \pi^*$ transition spectra of the supermolecular clusters of $CH_2O\text{–}nH_2O$ (n=1, 2, 3) are studied based on the structures optimized by quantum mechanics methods or extracted from molecular dynamics simulation. The results are consistent from our two models.

Chapter 4 - Formaldehyde (HCHO) is an important chemical feedstock and constituent of many industrial products, and is widely used in various adhesives and coatings of building materials. However, apart from its important application in chemical industry, HCHO is also defined as the most common and the best-known indoor air pollutant. Long time exposure to the indoor air with heavy HCHO pollution may cause serious health problems, such as irritation of the eyes, skin irritation, respiratory diseases, and even nasopharyngeal cancers. Thus, indoor air pollution has already aroused increasing concern, and great efforts have been made to eliminate HCHO pollution. Low temperature catalytic oxidation of HCHO is regarded as one of most attractive approach for elimination of HCHO, as HCHO can be completely converted into CO_2 and H_2O through catalytic oxidation process.

This paper is intended to review the recent advances in elimination of indoor air pollutant of HCHO through catalytic oxidation over oxide supported noble metal nanocatalysts. It is composed of four sections: (1) Overview of indoor air pollutant of HCHO including its chemistry, toxicology and source; (2) Various methods for elimination of HCHO pollution;

(3) Decontamination of HCHO pollution through catalytic oxidation over oxide supported noble metal nanocatalysts; (4) Summary and outlook.

Chapter 5 – The author describes their developed formaldehyde sensor element, monitoring device, and its application to indoor air quality measurement. The sensor element the authors developed is made of a porous glass that is impregnated with both ammonium ions and 1-phenyl-1,3-butandione. The color of the sensor elements changes from colorless to yellow after exposure to formaldehyde with a peak wavelength of 415 nm. There is a linear relationship between the 415 nm absorbance of the sensor elements and the accumulated formaldehyde concentration. The authors estimated the formation reaction rate constant of lutidine derivatives (yellow dye) on the sensor element, and also estimated quantity of interference gases. The authors found that the reaction occurred sufficiently quickly for them to monitor hourly changes in the formaldehyde concentration. The authors also found that there were no interference gases under normal atmospheric conditions. The developed sensor element was a small, flat plate, pumping-free, and accumulated type, therefore the authors could install it for an arbitrary period in a space whose formaldehyde concentration the authors wished to determine. Then the authors could convert the absorbance change of the sensor element into the formaldehyde concentration using a preliminarily calculated calibration curve.

The authors also developed a portable device for formaldehyde monitoring, and carried out indoor air monitoring in several houses. The absorbance difference of the developed sensor element was measured at regular intervals in the monitoring device and converted into the formaldehyde concentration. This was possible because the lutidine derivative that was formed as a yellow product of the reaction between β-diketone and formaldehyde was stable in the sensor element. The detection limit was 5 ppb x hour. The monitoring device is small and easy to use and the authors used it to perform hourly formaldehyde monitoring using their monitoring device under several indoor conditions. The authors found that a high formaldehyde concentration could be measured in a room containing furniture and clothes. The authors also found that, although the formaldehyde concentration decreased rapidly when the room was ventilated, it recovered rapidly in several hours when they stopped the ventilation.

Chapter 6 - Formaldehyde is of great importance in industry and in research on polymers. Formaldehyde was and still is an intense research topic in electrochemistry. Consequently, it has played a key role in fundamental studies on the complex mechanisms for the electrochemical oxidation of small-oxygenated organic molecules. Following a survey of studies on the electrochemistry of formaldehyde, electrochemical behaviors not that well known are discussed. The oscillatory potential in response to the applied current was monitored during the electrochemical oxidation of formaldehyde. A sequence of temporal states was found consisting of intervals of periodic and chaotic behaviors. In part of the range the oxidation of formaldehyde exhibits a sequence of period doubling bifurcations. Anions usually inhibit chemical reactions. It is shown under potential control conditions and specified other conditions that nitrate, which typically inhibits reactions, can enhance the electrochemical oxidation of formaldehyde.

In: Formaldehyde
Editors: Chan Bao Cheng and Feng Hu Ln

ISBN 978-1-62257-214-4
©2012 Nova Science Publishers, Inc.

Chapter 1

PROPERTIES OF UREA-FORMALDEHYDE RESINS FOR WOOD-BASED COMPOSITES

Byung-Dae Park
Department of Wood Science and Technology, Kyungpook National University, Daegu, Republic of Korea

ABSTRACT

This chapter reviews recent progresses on properties, chemical structure, thermal curing behavior, hydrolytic stability, morphology, microstructure, crystalline structure, and modifications of urea-formaldehyde (UF) resin as an adhesive for wood-based composite panels, particularly by focusing on the parameters related to formaldehyde emission (FE), such as synthesis reaction pH conditions, formaldehyde/urea (F/U) mole ratio, and resin modifications.

The reaction pH condition of UF resin synthesis showed that the amount of free formaldehyde strongly affected the reactivity of UF resin, and also indicated that the weak acid reaction condition provided a balance between increasing resin reactivity and improving adhesion strength of UF resins. Solid-state ^{13}C-NMR spectroscopy indicated that the molecular mobility of cured UF resin increased with decreasing the reaction pH used during its synthesis. The ^{13}C-NMR spectroscopy showed that UF resins with higher F/U mole ratios (i.e., 1.6 and 1.4) had two distinctive peaks, indicating the presence of dimethylene ether linkages and methylene glycols, which give a greater contribution to the FE than that of lower F/U mole ratio. However, these peaks were not detected at the UF resins with lower F/U mole ratios (i.e., 1.2 and 1.0). Lowering F/U mole ratio of UF resins as a way of abating FE consequently requires improving their reactivity. As the F/U mole ratio decreases, thermal curing behavior of these UF resins such as the gel time, onset and peak temperatures, and heat of reaction (ΔH) increased, while the activation energy (Ea) and rate constant (k) were decreased. The results also suggested that as the F/U mole ratio decreased, the FE of particleboard (PB) was greatly reduced at the expense of the reactivity of UF resin and slight deterioration of performance of PB prepared. Dynamic mechanical analysis (DMA) results partially explained the reason why UF resin adhesives with lower F/U mole ratio resulted in relatively poor adhesion performance.

Morphological investigation on UF resins illustrated that the spherical structures in cured UF resins were much more resistant to the hydrolytic degradation by the acid than amorphous region. Atomic force microscopy (AFM) images showed two distinctive regions, i.e., hard and soft phases in cured UF resins. The AFM study suggested that the soft phase was much more susceptible to the hydrolysis of cured UF resin than the hard phase. The soft phase of cured UF resins by ammonium chloride was much more easily hydrolyzed than those cured by ammonium sulfate, indicating that hardener types had a great impact on the hydrolytic degradation behavior of cured UF resins. For the first time, the presence of thin filament-like crystalline structures on the fracture surface of cured UF resin was reported. And X-ray diffraction (XRD) results showed that the crystalline regions of cured UF resins with lower F/U mole ratio contribute partially to the improved hydrolytic stability of the cured resin.

1. INTRODUCTION

Urea-formaldehyde (UF) resin is a polymeric condensation product of the chemical reaction of formaldehyde with urea, and is most widely used as adhesive for manufacturing of wood-based composite panel, particularly plywood, particleboard or medium density fiberboard. Therefore, UF resin adhesives are considered as one of the most important wood adhesives. So, wood panel industry is a major user of UF resin adhesives. For example, the North America's production of formaldehyde-based resin in 1999 was 3.3 million tons, of which 56.6% is UF resins, and 40.3% is PF resins [1]. The production of wood adhesive in European countries including was 5.1 million tons in 2003, of which 69.6% was UF resins [2]. In China, about 1.8 million tons of wood adhesives were produced, and about 63.4% was UF resins in 2003 [3].

UF resin adhesive possesses some advantages such as fast curing, good performance in the panel, water solubility and lower price. Disadvantages of using the UF resin are formaldehyde emission (FE) from the panels and lower resistance to water. Lower resistance to water limits the use of wood-based panels bonded with UF resin to interior applications. However, the FE from the panels used for interior applications was one of the factors, affecting sick building syndrome in indoor environment.

Free formaldehyde present in UF resin and hydrolytic degradation of UF resin under moisture condition has been known as responsible for the FE from wood-based panels [4]. In other words, un-reacted formaldehyde in UF resin after its synthesis could be emitted from wood panels even after hot-pressing at high temperature. In addition, the reversibility of the aminomethylene link and its susceptibility to hydrolysis also explains lower resistance against the influences of water and moisture, and subsequently FE [5]. Therefore, the FE issue has been one of the most important aspects of UF resin in last few decades [6-11].

Much attention has been paid to reduce or control the FE from UF resin-bonded panels through resin technologies. Until the mid-sixties, most UF resins were synthesized by the two-step reaction procedures: i.e., methylolation and condensation. In other words, the methylolation reaction was done under alkaline condition followed by the condensation reaction under acidic condition [12]. This synthesis method was widely employed for UF resin preparations for a long time. In the early seventies, however, this method faced the serious problem of the FE. So, lowering the formaldehyde to urea (F/U) mole ratio for the synthesis of UF resin was adopted as one of the approaches to reduce the FE of UF resin-bonded panels [9]. Thus, lower F/U mole ratios from 1.1 to 1.2 started to be used for the resin synthesis.

An excellent literature review on the influence of F/U mole ratio on the FE as well as panel properties has been done by Myers [13]. According to the review, the gel time used as an indicator of resin reactivity increased with decreasing F/U mole ratio. In general, lower F/U mole ratios cause less FE form the panel with a loss of panel properties, particularly internal bond (IB) strength as well as thickness swelling after water immersion for 24 hours. Lower F/U mole ratios also reduced modulus of rupture (MOR) [9]. In recent years, it was reported that close F/U mole ratios produced quite similar structures and performance in UF resin, leading to the conclusion that the most important factor in synthesis of UF resin was the F/U mole ratio [14]. This chapter introduces recent progresses on properties and FE of UF resins as adhesives for manufacturing wood-based composite panels. Firstly, this chapter attempts to provide chemistry of UF resins, the effects of synthesis parameters on the properties of UF resins.

2. Chemistry of UF Resins

The use of different conditions of reaction and preparation could produce a broad variety of UF resins. Basically, the reaction of urea and formaldehyde is a two-step process: usually alkaline methylolation followed by an acid condensation. The combination of these two chemicals results in linear and/or branched as well as tridimensional network in the cured resin. This is due to the functionality of four in urea (due to the four replaceable hydrogen atoms), and that of two in formaldehyde. The most important factors determining the properties of the reaction products are: 1) the relative molar proportion of urea to formaldehyde, 2) the reaction temperature and time, and 3) the various pH values at which the condensation takes place [15].

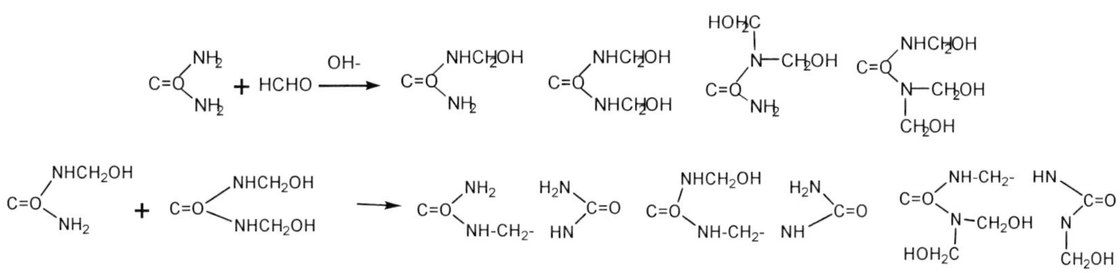

Figure 1. Methylolation and condensation reactions in UF resins.

Figure 1 illustrates the methylolation and condensation reactions in UF resin. The alkaline condensation (i.e., methylolation) refers to the addition of up to three (four in theory) molecules of the bifunctional formaldehyde to one molecule of urea to give the so-called methylolureas. The molecular species of the methylolation are mono-, di-, and trimethylolureas as shown in Figure 1. It is known that tetramethylolurea has never been isolated [15]. Low temperature and weak acidic pH favors the formation of methylene ether bridges ($-CH_2-O-CH_2-$) over methylolation [16]. Each methylolation step has its own rate constant (k), with different k values for the forward and backward reactions. The reversibility of this reaction is one of the most important aspects of UF resins. This feature is responsible for both the low resistance against hydrolysis and the subsequent FE. An acid condition for UF resin synthesis is known to produce varieties of uronic derivatives [16]. The presence of some of these species was detected by many other studies [11, 17-19].

Many authors investigated the chemical structures of UF resins using ^{13}C-NMR spectroscopy to understand their reaction mechanisms, and chemical constitutions. To the authors' knowledge, Ebdon and Heaton [20] have done the first work on the chemical structure of UF resin, showing that the ^{13}C-NMR spectroscopy was useful in providing information about the chemical constitution of UF resin. In the following year, Tomita and Hatono [19] did an intensive ^{13}C-NMR work on UF resin by assigning the chemical shifts and quantifying the quantity of specific chemical species with a particular structure. Meyer [21] also employed the same tool to relate chemical structures of UF resin to the FE issue, and mentioned that the formaldehyde release was due to the weakest chemical links in the resin like ether, or pendant methylol groups. Kim and Amos [22] also used the same tool to investigate the influence of initial F/U mole ratios with a fixed final F/U mole ratio of 1.0 to the chemical structures of UF resin prepared. They reported that the emitted formaldehyde level of particleboard decreased with a decrease in the initial F/U mole ratio. Gu *et al.* [11] studied the chemical structures of UF resins prepared under different pH conditions, and reported that the content of branched structures increased as the pH decreased. Christjanson *et al.* [14] applied the same tool for the investigation of structural changes of UF resins during storage, and showed that the main reaction during storage was the formation of methylene linkages.

A series of very extensive work on the chemical structures of UF resins using the ^{13}C-NMR spectroscopy was done by Kim's group [23-27]. Kim [23] reported that monomeric methylolureas and methyl-ether derivatives were formed in the initial alkaline reaction while methylene bonds were formed in the subsequent acidic reaction, splitting formaldehyde from methylene-ether linkages. Kim [24] also showed that the addition of the second urea during UF resin synthesis led various polymeric methylolated ureas migrate to monomeric methylolated ureas. Kim [25] also found a decreasing number of side chain branches and increasing number of free amid group as the initial F/U mole ratio decreased from 2.4 to 1.8 with the final F/U mole ratio of 1.15. Kim *et al.* [25-27] reported the relationship between the chemical structures of UF resins and the FE of particleboard, depending on post-treatments and initial F/U mole ratios. Ferg *et al.* [28] showed that a quantitative measurement of peak ratios could be used to predict UF resin strength and subsequent FE of the resultant particleboard. Tohmura *et al.* [18] have studied the change of chemical structure of UF resins prepared under different reaction pH conditions using ^{13}C-NMR technique. One of the findings of these studies was the detection of uronic structures in the UF resin prepared under a strong acid condition.

In addition, the influence F/U mole ratio to the chemical structure of UF resins has also been done. Figure 2 shows the ^{13}C-NMR spectra of the UF resins with different F/U mole ratios. The assignments of chemical structures from the ^{13}C-NMR spectra are also summarized in Table 1. The ^{13}C-NMR spectra of UF resins with different F/U mole ratios showed a peak at around 45 ppm. This chemical shift around 45 ppm was assigned to the carbon of methylene linkages, according to published information [10, 17-18, 23, 27]. The chemical shift of around 54 ppm could be attributed to the carbons of methylene linkage such as dimethylene urea, trimethylene tetraurea, or tetramethylene pentaurea [19]. But, these methylene linkages were not detected for the UF resin with the F/U mole ratio of 1.0. This result indicates that the chemical structure of UF resin with the F/U mole ratio was much less branched polymer without trimethylene tetraureas, or tetramethylene pentaureas. These species were known to contribute the formation of branched network polymer structures [29].

The chemical shift at around 63 ppm was assigned to various methylol carbons of the resins with different F/U mole ratios. The chemical shifts could be due to the carbons of monomethylol ureas, or dimethylol ureas. This result was well in the agreement with the published report [11]. The dimethylene ether carbon occurred at around 68 ppm for the UF resins with the F/U mole ratios of both 1.6 and 1.4. In other words, this peak was fairly weak for the resins prepared at the F/U mole ratios of 1.2 and 1.0. This result indicates that UF resins with higher F/U mole ratios tend to form dimethylene ether linkages. These linkages were also reported by other published papers [19, 25, 30, 31]. These structures were known to susceptible to hydrolytic degradation under acidic environment [30].

The chemical shift of around 81.8 ppm was occurred for UF resins with the F/U mole ratios of 1.6 and 1.4 (arrows in Figure 2 a, b). However, this was not detected for the UF resins with the F/U mole ratios of 1.2 and 1.0. This chemical shift was assigned methylene glycols that were a dissolved form of free formaldehyde in the resin. Chung and Maciel [30] reported the chemical shift of 87-92 for the methylene glycols while Gu *et al.* [31] reported the range of from 83 ppm to 95 ppm as the methylene glycol species. This result shows that higher F/U mole ratio produces more free formaldehyde than lower F/U mole ratio for the synthesis of UF resins.

Table 1. Chemical shift assignment of ^{13}C-NMR spectra of UF resins synthesized under different reaction pH conditions

Chemical structure	Chemical shift (ppm)			
	F/U = 1.6	F/U =1.4	F/U =1.2	F/U =1.0
$NH\text{-}\underline{C}H_2\text{-}NH$	45.26	45.23	45.27	45.40
$N(CH_2\text{–})\underline{C}H_2N(CH_2\text{–})\text{–}$	54.19	54.14	54.18	
$NH\text{-}\underline{C}H_2OH$	63.39	63.39	63.39	63.41
$N(CH_2\text{-})\underline{C}H_2OH$	63.97	63.54	63.54	63.56
$\text{-}NH\text{-}\underline{C}H_2\text{-}O\text{-}\underline{C}H_2\text{-}NH\text{-}$	68.91	68.91		
$HO\text{-}\underline{C}H_2\text{-}OH$	81.89	81.88		
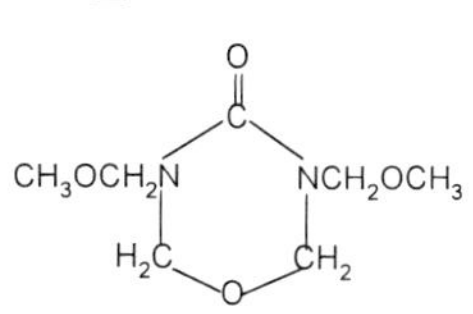 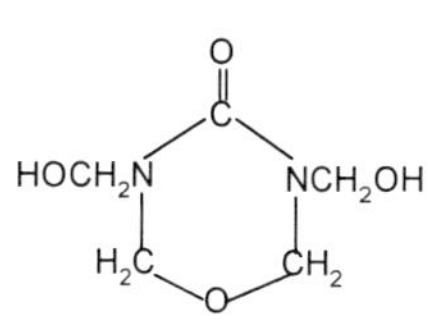	156.75 156.90	156.89	156.89	156.94
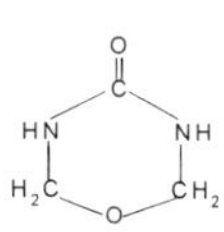	157.15		157.36	
$HOCH_2NH\underline{C}ON(CH_2OH)_2$	158.07	158.06	158.06	158.12
$\text{-}HN\text{–}\underline{C}O\text{–}NH\text{-}, NH\text{–}\underline{C}O\text{–}NH_2\text{–}$		159.44	159.44	159.51

Reproduced from ref. [69] with by permission from Mokchae Konghak (© Mokchae Konghak, 2008).

The chemical shifts around 156 ppm could be assigned to the carbonyl carbons of uronic structures [11, 31, 20]. The peak at 156.7 ppm could be assigned to the carbonyl carbons of urons with the di-substitutions of dimethylene methyl ether groups (CH_3OCH_2N-C-O-NCH_3OCH_2), while the one at 156.9 could be assigned to the carbonyl carbons of urons with the di-substitutions of dimethylol groups ($HOCH_2N$-C-O-NCH_2OH). And the chemical shift at around 157 ppm was assigned to the carbonyl carbons of uronic structures without any substitutions [11, 31, 20]. This assignment is in a good agreement with the published results. Three peaks at 158 ppm and 159 ppm were appeared prominently for all four UF resins. The peak at around 158 ppm was assigned to the carbons of carbonyl group of tri-methylolated ureas [11, 30, 31]. The one at around 159 ppm was assigned to the carbons of substituted carbonyl groups of ureas [17-19]. The peak at 159 ppm was assigned to the carbons of carbonyl group of various urea residues [11, 31].

Table 2. Absorption band assignments of FT-IR spectra of UF resins synthesized under different reaction pH conditions

Absorption band (cm^{-1})	Chemical structure assignment	Observed band (cm^{-1}) pH: 7.5	pH: 4.5	pH: 1.0
3350-3340	NH stretching of primary aliphatic amines	3340	3349	3349
2962-2960	–O–CH_3, aliphatic ethers	2962	2959	2962
1654-1646	C=O stretching of primary amide	1653	1646	1654
1560-1550	C–N stretching of secondary amines	1559	1560	1560
1465-1440	C–H bending in NCH_2N, CH_2O, OCH_3	-	-	1465
1400-1380	C–H mode in CH_2 and CH_3	1387	1391	1390
1380-1330	C–N stretching of CH_2–N	1349	1354	-
1320-1300	=C–N or =CH–N of tertiary cyclic amides	-	-	1319
1260-1250	C–N and N–H stretching of tertiary amides	1253	1256	1259
1150-1130	C–O stretching of aliphatic ether	1133	1134	1133
1050-1030	C–N or NCN stretching of methylene linkages (NCH_2N)	1032	-	1024
1020-1000	C–O stretching of methylol group	-	1001	-
900-650	N–H bending of primary aliphatic amines	776	780	805
750-700	N–H bending of secondary aliphatic amines (R1-CH_2-NH-CH_2-R2)	-	-	752

Reproduced from ref. [36] with by permission from John Wiley & Sons (© John Wiley & Sons, 2003).

In addition to the F/U mole ratio, chemical structures of UF resins prepared under synthesis pH conditions using both fourier transform infrared (FT-IR) and ^{13}C-NMR spectroscopy-was also used. The FT-IR spectra of UF resins prepared under three different reaction conditions are shown in Figure 3. Also, the assignments of chemical structures from FT-IR spectra are summarized in Table 2. The band at 1465 cm^{-1} was only detected for the

UF resin prepared under the strong acid reaction condition. This might have resulted from the presence of uronic derivatives in the resin as supported by the result of ^{13}C-NMR spectrum of the same resin (Table 3). And, this result was also compatible with the occurrence of 1320 cm^{-1} band assigned to =C-N or -CH-N structure as reported [33]. The spectra of both resins from alkaline and strong acid reaction conditions showed strong bands at 1032 cm^{-1} and 1024 cm^{-1}, respectively. The band from 1030 to 1050 cm^{-1} could be assigned to C-N stretching from C-N or C-N$_2$, of amides in the resins [Myers 1981]. So, these bands could result from the methylolated ureas or methylene ureas. However, this region was relatively weak for the resin prepared under weak acid condition. Instead the resin from weak acid condition showed 1001 cm^{-1} band, which was assigned to the C-O stretching of methylol groups [34]. This result suggests that methylolated and methylene ureas are present for all the resins from three different reaction conditions. But, the infrared spectra suggest that types of these chemical species present in the resins were quite different from each other. The infrared spectra and band assignments to chemical structures produced useful information on the resins synthesized under different reaction pH conditions. The C=O stretching of primary amides was most dominant absorption band of infrared spectra of UF resin which occurred at 1650 cm^{-1}. Two bands at 1465 cm^{-1} and 1320 cm^{-1} supported the presence of uronic derivatives in the UF resin prepared under strong acid reaction condition.

Table 3. Chemical shift assignment of ^{13}C-NMR spectra of UF resins synthesized under different reaction pH conditions

Chemical shift (ppm)	Chemical structure	Observed chemical shift (ppm)		
		pH: 7.5	pH: 4.5	pH: 1.0
44-45	NH-$\underline{C}H_2$-NH	44.9	44.8	45.2
53-54	N(CH_2–)$\underline{CH_2}$N(CH_2–)–	53.6 - 54.1	53.7	-
62.9	NH-$\underline{C}H_2$OH	62.9	62.9	-
67.0	-NH-$\underline{C}H_2$-O-$\underline{C}H_2$-NH-	-	67.0	-
71.7	N(CH2-)$\underline{C}H_2$OH	71.3	71.3	-
73.5-75.2	uron -$\underline{C}H_2$-O-$\underline{C}H_2$-uron	-	-	74.0
78.8	HN-CO-NH | | $H_2\underline{C}$-O-$\underline{C}H_2$	-	-	78.4
155.1	HN-$\underline{C}$O-NH | | H_2C-O-CH_2	-	-	156.0
157.2-157.7	-HN–$\underline{C}$O–NH-	157.1-157.9	157.0-157.9	157.9
161	-NH–$\underline{C}$O–NH_2–	158.2	158.2	158.6
159-163	$HOCH_2NH\underline{C}ON(CH_2OH)_2$	159.2	159.2	159.7
166.3	H$\underline{C}$OOH	166.3	-	-

Reproduced from ref. [36] with by permission from John Wiley & Sons (© John Wiley & Sons, 2003).

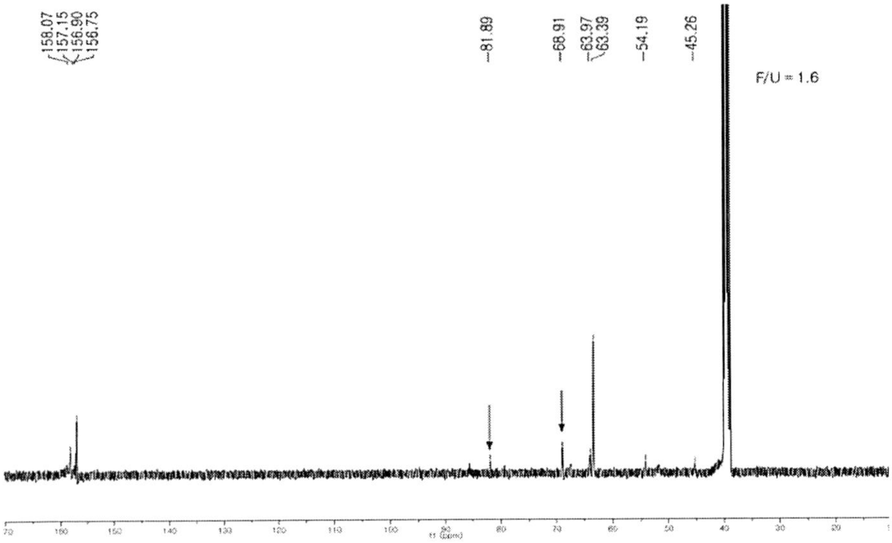
158.07
157.15
156.90
156.75
81.89
68.91
63.97
63.39
54.19
45.26
F/U = 1.6
f1 (ppm)

(a)

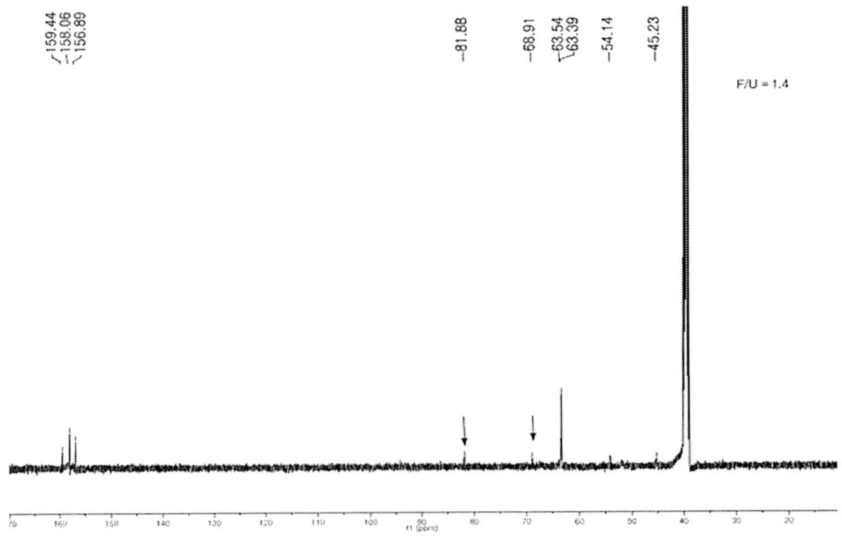
159.44
158.06
156.89
81.88
68.91
63.54
63.39
54.14
45.23
F/U = 1.4
f1 (ppm)

(b)

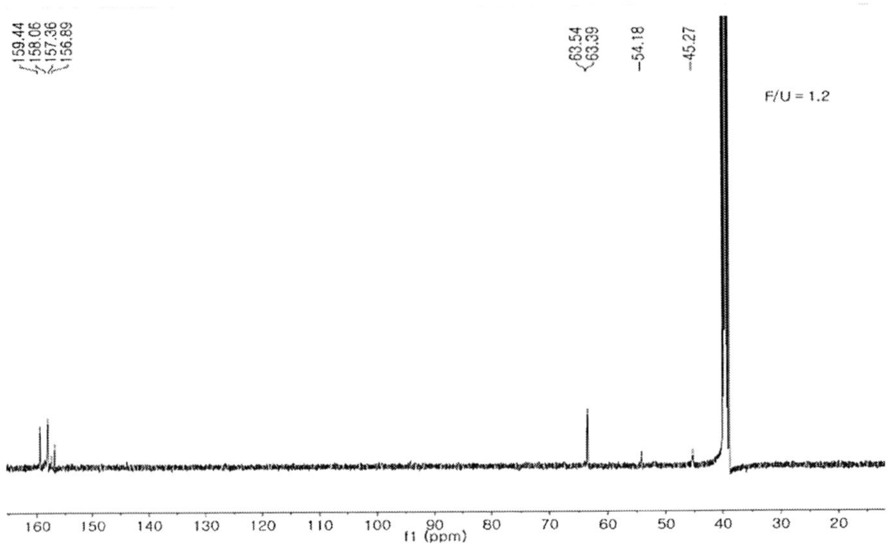
159.44
158.06
157.36
156.89
63.54
63.39
54.18
45.27
F/U = 1.2
160 150 140 130 120 110 100 90 80 70 60 50 40 30 20
f1 (ppm)

(c)

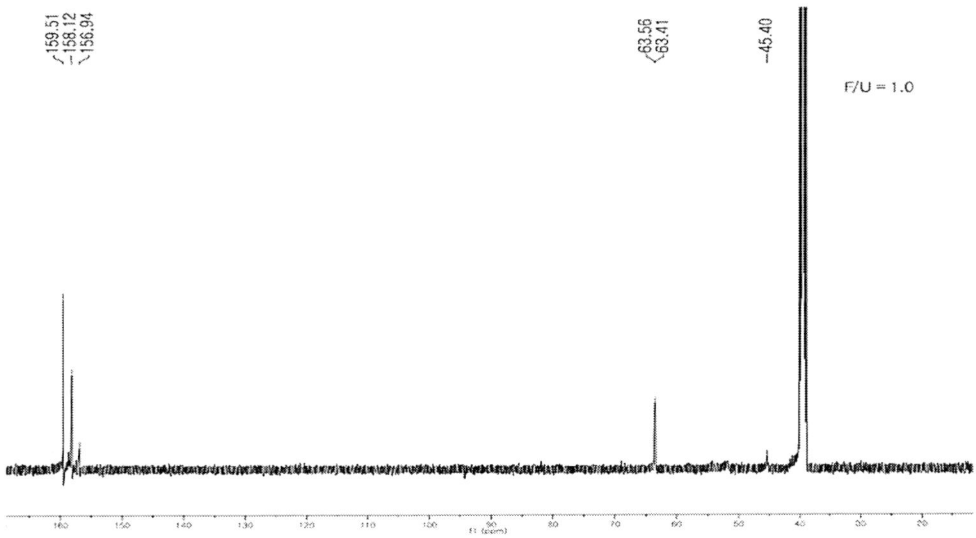

Figure 2. ^{13}C-NMR spectra of neat UF resins with different F/U mole ratios. (a) 1.6, (b) 1.4, (c) 1.2, and (c) 1.0.
* Reproduced from ref. [69] with by permission from Mokchae Konghak (©Mokchae Konghak, 2008).

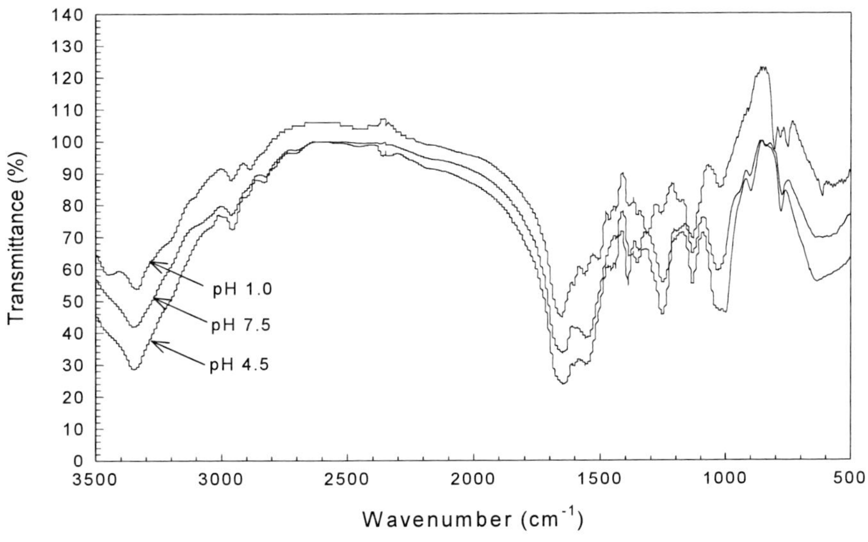

Figure 3. FT-IR spectra of UF resins prepared under three different reaction pH conditions.
*Reproduced from ref. [36] with by permission from John Wiley & Sons (© John Wiley & Sons, 2003).

Table 3 shows the chemical shift and structural assignments of ^{13}C-NMR spectra of UF resins synthesized under different reaction pH conditions. In general, the spectra of two UF resins synthesized under weak acid and alkaline reaction conditions were similar to each other while the spectrum of the resin prepared under strong acid reaction condition was different. Both UF resins produced under alkaline and weak acid reaction conditions showed a peak at 44 ppm while the resin synthesized under strong acid reaction condition had a peak at 45 ppm. The chemical shift from 44 to 45 ppm was assigned to the carbon of methylene linkages, according to published information [10-11, 17-19, 22]. The chemical shift from 53 to 54 ppm might be attributed to various methylene carbons such as dimethylene urea, trimethylene tetraurea, or tetramethylene pentaurea [19]. But, these methylene linkages detected were extremely small for the resin prepared under strong acid condition.

The chemical shift at 62-63 ppm was assigned to various methylol carbons of the resins prepared under alkaline and weak acid conditions. This peak for the resin synthesized under strong acid condition was relatively weak compared to other two reaction conditions. The methyl ether carbon occurred at 67 ppm for the weak acid resin. This peak was fairly weak for the resin from both alkaline and strong acid conditions. The chemical shift of 71.3 ppm from the spectra of both alkaline and weak acid reaction pH resins was assigned to substituted methylol carbons [17, 19]. However, Gu *et al.* [11] and Tohmura *et al.* [18] assigned the substituted methylol carbons to 72 ppm, which was shifted to downfield compared with 71 ppm. These results explain why the UF resin synthesized under strong acid condition had longer gel time that those of alkaline and weak-acid resins.

The chemical shift of 74 ppm from the spectrum of the strong acid resin might be assigned to methylene carbons of uron structure. As mentioned, uronic structures were found for the resin prepared under strong acid condition. Hence, this assignment is in a good agreement with the published results. Other carbons from uronic structures were also detected 153.3 ppm for the strong acid reaction resin. These chemical shifts were also assigned to the carbonyl group of the uronic structure [11, 17]. The chemical shift of 78.4 ppm was also strongly detected for the strong acid resin, and was assigned to methyl ether of methylol group [19].

Three peaks at 157, 158, and 159 ppm were appeared prominently for all three resins. The peak at around 157 ppm was assigned to the carbons of carbonyl group of urea [11, 18]. The chemical shift of 158 ppm was assigned to the carbons of substituted carbonyl groups [11, 17-18]. The peak at 159 ppm was assigned to the carbons of carbonyl group of various urea residues [11]. The chemical shift of 166 ppm was supposed to be due to formic acid used for the pH control during resin synthesis [17].

Table 4. Chemical shift assignments of solid-state ^{13}C CP/MAS NMR spectra of UF resins synthesized under different reaction pH conditions

Chemical shift (ppm)	Chemical structure	Observed chemical shift (ppm)		
		pH: 7.5	pH: 4.5	pH: 1.0
44-45	$NH\text{-}\underline{C}H_2\text{-}NH$	46.9	46.9	46.4
53-54	$N(CH_2\text{-})\underline{C}H_2N(CH_2\text{-})\text{-}$	54.5	54.5	55.1
62.9	$NH\text{-}\underline{C}H_2OH$	64.4	64.9	-
67.0	$\text{-}NH\text{-}\underline{C}H_2\text{-}O\text{-}\underline{C}H_2\text{-}NH\text{-}$	68.5	-	-
71.7	$N(CH_2\text{-})\underline{C}H_2OH$	73.0	72.1	-
78.8	Uron $CH_2\text{-}O\text{-}CH_2$	-	-	78.3
155.1	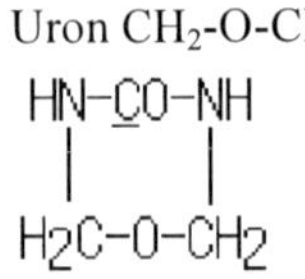	-	-	155.9
159-163	$HOCH_2NH\underline{C}ON(CH_2OH)_2$	159.7	159.9	160.6

Reproduced from ref. [36] with by permission from John Wiley & Sons (© John Wiley & Sons, 2003).

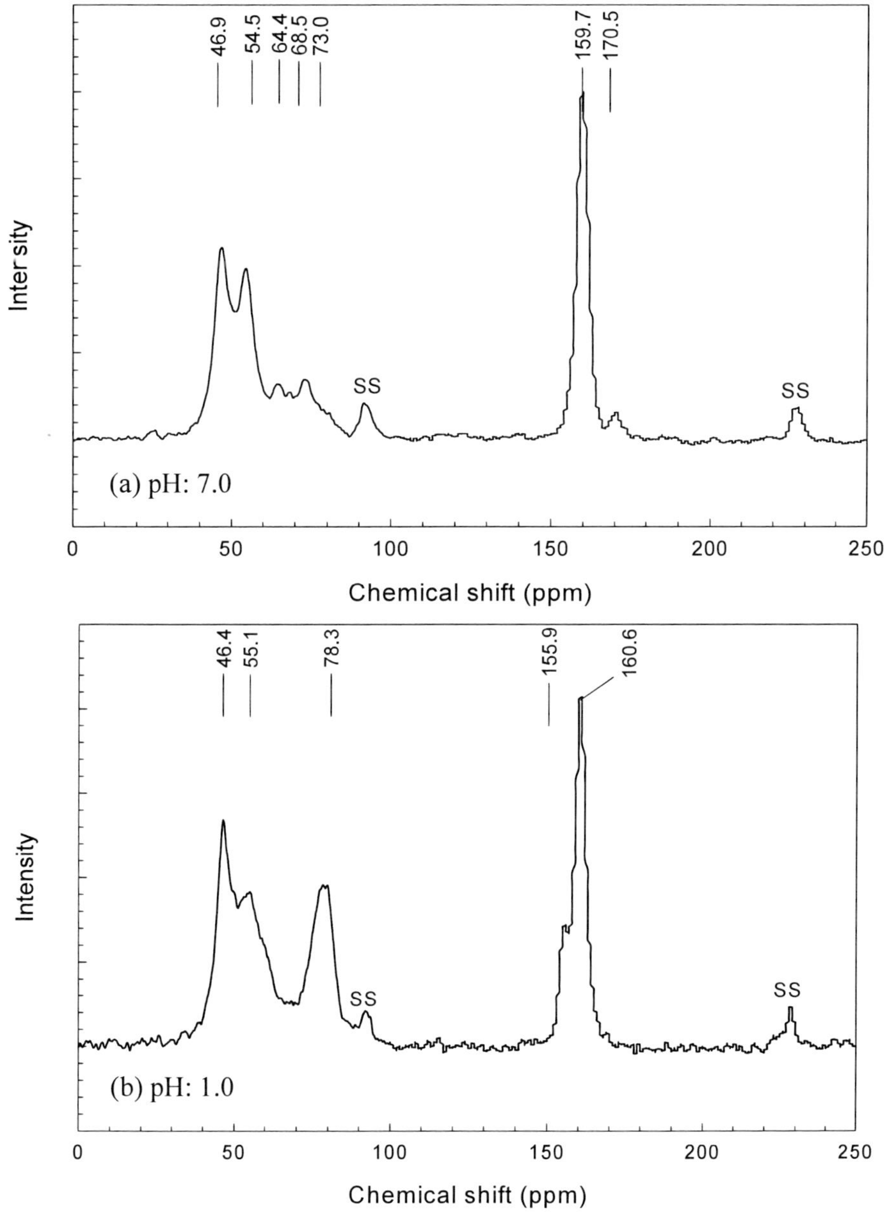

Figure 4. ^{13}C CP/MAS NMR spectra of UF resins prepared under (a) alkaline and (b) strong acid reaction condition. SS means side spinning peak.
*Reproduced from ref. [36] with by permission from John Wiley & Sons (© John Wiley & Sons, 2003).

Typical spectra of solid-state ^{13}C CP/MAS NMR spectroscopy for UF resins prepared under pH levels are shown in Figure 4. The peak assignments for those spectra were summarized in Table 4. Three UF resins synthesized under three different conditions had a peak in common at 46 ppm, which indicates the presence of methylene linkages ($-CH_2-$) in UF resin. The peaks from 54-55 ppm assigned to methylene linkages were also present for all three spectra of the UF resins. This result was in agreement with other's work [18, 35].

However, the intensity of this peak was relatively large compared with the one from liquid ^{13}C NMR spectrum of the UF resin prepared under strong-acid condition. This might be solvent effect for liquid ^{13}C NMR technique. The chemical shift of 64 ppm was assigned to methylol carbon of methylolated ureas. The peak was not strong for the UF resin from the strong-acid reaction condition, which was similar to the liquid ^{13}C NMR spectrum. This result again indicates that the strong-acid reaction condition is not favorable to from methylol groups (CH_2OH) in the UF resin.

The spectra from UF resins from both the alkaline and weak acid condition did not show the peak at 78 ppm, indicating that uronic structures were not present in both resins (Figure 4, a). One of the distinctive peaks from the UF resin synthesized under strong acid condition was the peak at 78 ppm, which was assigned to the carbons of CH_2 groups of uronic derivatives. Soulard *et al.* [17] reported that the presence of uronic structures improves the adhesion strength of UF resins. This result is also compatible with other published results [27-28]. Thus, this result suggested that the strong acid condition might contribute to improve adhesion strength of UF resin. And the peak at 155 ppm was also assigned to the carbonyl group of uron [17, 19]. The peak at 159-160 ppm was assigned to the carbonyl group of urea as well as methylolated ureas.

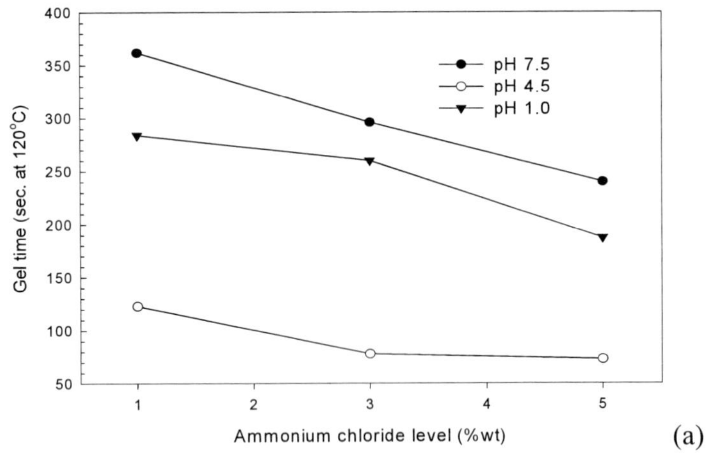

(a)

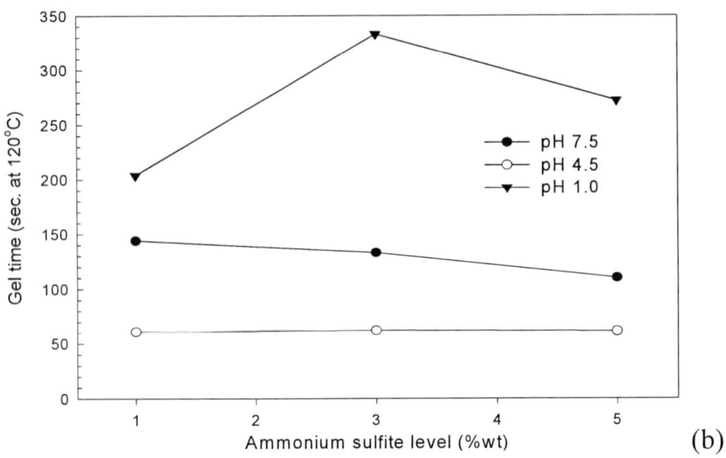

(b)

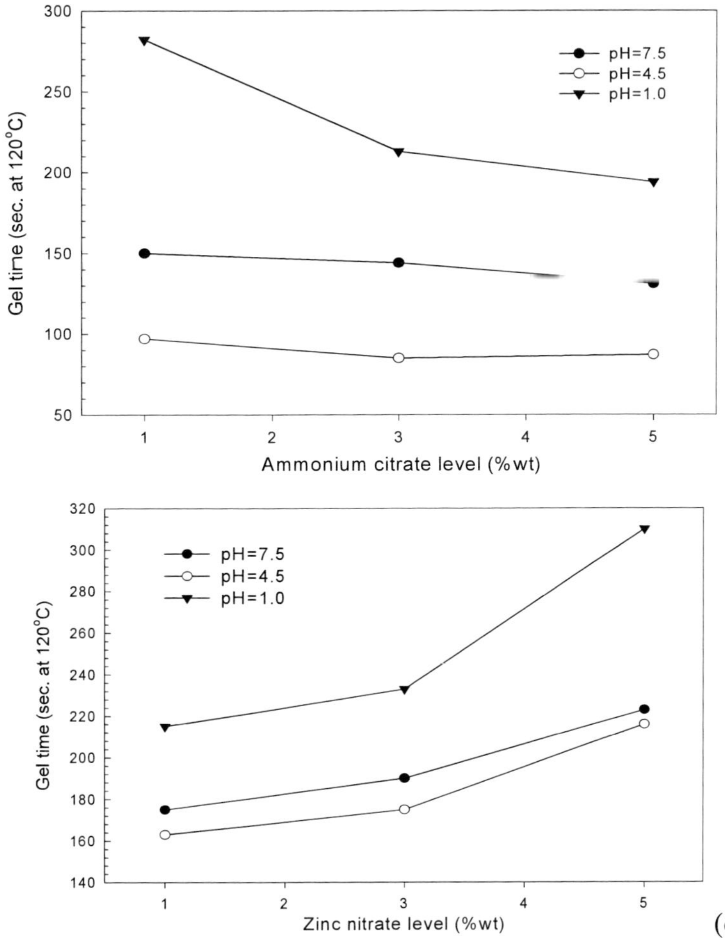

Figure 5. Gel times of UF resins synthesized under three different reaction pH conditions, hardener type and level. (a) Ammonium chloride, (b) ammonium sulfate, (c) ammonium citrate, and (d) zinc nitrate. *Reproduced from ref. [36] with by permission from John Wiley & Sons (© John Wiley & Sons, 2003).

The results of the proton spin-lattice rotating frame relaxation times ($T_{1\rho H}$) were summarized in Table 5. The $T_{1\rho H}$ value of the peak at 46 ppm assigned to methylene linkages decreased up weak acid (pH: 4.5) and then slightly increased for strong acid (pH: 1.0) condition. Traditional UF resin synthesis include methylolation at alkaline condition usually at the pH range from 5.0 to 8.0 and then condensation at acid condition at the pH range from 3.0 to 5.0. It was known that more branched polymers are formed in methylolation reaction while methylene linkages are formed in condensation reaction [16]. In other words, alkaline condition would produce short and sterically hindered polymers while acid condition would produce less cross-linked and pliable polymers. Thus, the results of $T_{1\rho H}$ measurement indicate that a greater relaxation time of the UF resin synthsized under alkaline condition gives rigid structures which do not easily relax. By contrast, a smaller relaxation time of the UF resin synthesized under strong acid condition indicated a greater mobility of chemical structure. In general, the $T_{1\rho H}$ value decreased with decreasing the reaction pH for UF resin synthesis. This result indicates that molecular mobility of UF resin increases with decreasing the reaction pH used during its synthesis.

Table 5. The proton spin-lattice rotating frame relaxation time ($T_{1\rho H}$) of UF resins prepared under three different reaction pH conditions using ^{13}C CP/MAS NMR spectroscopy

Peak (ppm)	$T_{1\rho H}$ (ms)		
	pH:7.5	pH: 4.5	pH: 1.0
160	27.7	15.3	14.3
54	15.4	6.8	5.9
46	14.6	9.9	13.1

Reproduced from ref. [36] with by permission from John Wiley & Sons (© John Wiley & Sons, 2003).

Table 6. Properties of UF resins synthesized at different pH levels

Reaction pH	Non-volatile solids content (%)	Initial F/U molar ratio	Final F/U molar ratio	Viscosity (cps)	Free HCHO (%)	Final pH
7.5	48.5	2.2	1.15	58	0.45	8.0
4.5	52.1	2.2	1.15	135	0.69	8.0
1.0	49.8	3.0	1.15	66	0.55	8.0

Reproduced from ref. [36] with by permission from John Wiley & Sons (© John Wiley & Sons, 2003).

3. Properties of UF Resins

3.1. General Properties of UF Resins

As discussed in the previous section, the reaction pH and F/U mole ratio have a great impact on the chemical structure of UF resins, which consequently influence properties of UF resins. For example, the properties of UF resins prepared under three different reaction conditions are summarized in Table 6. The resin prepared under weak acid condition showed relatively greater viscosity and higher free formaldehyde compared with other two resins prepared. The non-volatile solids content of UF resins prepared under three different reaction conditions were about 50% by weight. The resin prepared under weak acid condition showed slightly higher viscosity and free formaldehyde content compared with other two resins prepared. The results of gel time measurements of the UF resins prepared under three different reaction conditions are shown in Figure 5. The gel time of all UF resins decreased with increasing ammonium chloride level (Figure 5, a). In spite of this, the UF resins synthesized under alkaline reaction condition showed the longest gel time at all hardener levels, followed by the strong acid and weak acid condition. However, the gel time of UF resins synthesized under weak acid condition was much shorter than that of the resins prepared under both alkaline and strong acid conditions. This result indicated that the weak acid reaction condition provide a better reactivity with UF resin compared with other reaction conditions. It is interesting that the gel time was not proportionately decreased when the hardener level was increased from 3% to 5%.

Figure 5, (b) presents the result of the gel time measurement of UF resins as a function of ammonium sulfate level. The gel time of two UF resins prepared alkaline and weak acid conditions decreased with increasing the hardener level. The gel time of the UF resin of strong acid condition increased with increasing the hardener level reaching a maximum at 3% ammonium sulfate level. Among three UF resins, the UF resin from weak acid condition showed the shortest gel time, indicating that the combination of weak acid condition with ammonium sulfate could provide a way of cure acceleration of UF resin.

The result of the gel time measurement of UF resins as a function of ammonium citrate level is shown in Figure 5, (c). As expected, the gel time decreased with increasing the hardener level. However, the resin prepared under the strong acid condition produced the longest gel time, followed by the alkaline and then weak acid condition. The result indicates that the resin prepared under weak acid condition had the reactivity when ammonium citrate was used. As a hardener, zinc nitrate was also used for the resins prepared under different reaction pH conditions. The gel time of the resins increased as the zinc nitrate level increased (Figure 5, d). In other words, zinc nitrate was not effective for UF resin as a hardener. The resin synthesized under weak acid condition showed the shortest gel time when ammonium sulfate was used as hardener.

In terms of the reactivity of UF resin for three different reaction conditions, the weak acid reaction condition produced much faster reactivity compared to alkaline and strong acid reaction conditions. Among four hardeners used, the ammonium sulfate gave much shorter gel time than other hardeners used. This result might be attributed to the extent of acidic nature of hardeners used. The more acidic hardener is the faster cure of UF resin is at the same temperature. Thus, the weak acid reaction condition would be a possible solution to accelerate the reactivity of UF resin.

The properties of UF resins prepared at different F/U mole ratio are summarized in Table 7. The non-volatile solids contents of UF resins prepared at different F/U mole ratios were ranged from about 50 to 54 wt%. The resin viscosity and specific gravity was not much different for the F/U mole ratios.

Table 7. Formaldehyde emissions and internal bond strength of particleboard bonded with UF resins prepared under three different reaction pH conditions

Reaction pH	Formaldehyde emission (mg/L)	Internal bond strength (MPa)
7.5	1.83	0.61
4.5	2.15	0.55
1.0	0.78	0.04

Reproduced from ref. [36] with by permission from John Wiley & Sons (© John Wiley & Sons, 2003).

3.2. Thermal Curing Behaviors of UF Resins

Figure 6 illustrates a typical DSC curve of UF resin prepared under weak acid condition, showing onset temperature, heat of reaction (ΔH), and peak temperature (T_p). The onset point is defined as the extrapolated beginning point of any transition or phase change determined from data analysis. Thus, the onset temperature (arrowed) may be expressed as an extrapolated and starting temperature of curing of the UF resin. The onset temperature of a UF resin for each heating rate was obtained from an average of three dynamic scans.

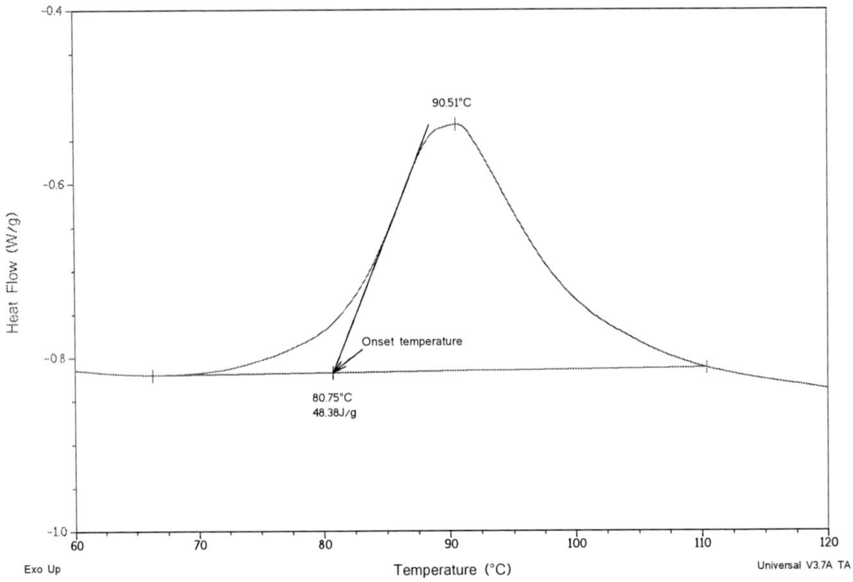

Figure 6. Typical thermogram of UF resin prepared under weak acid condition, showing the onset temperature and heat of reaction (ΔH) at the heating rate of 5°C/min.
*Reproduced from ref. [60] with by permission from John Wiley & Sons (© John Wiley & Sons, 2006).

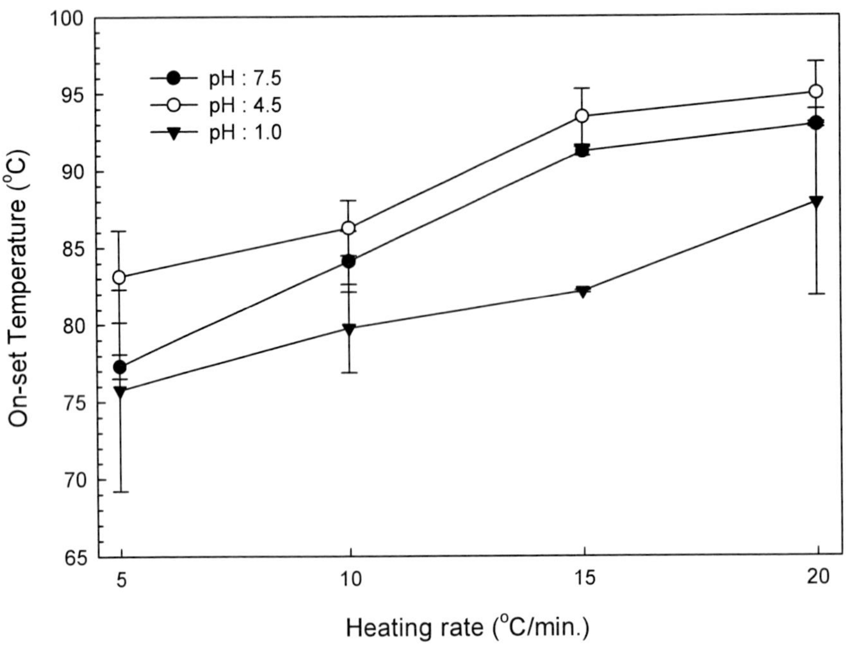

(a)

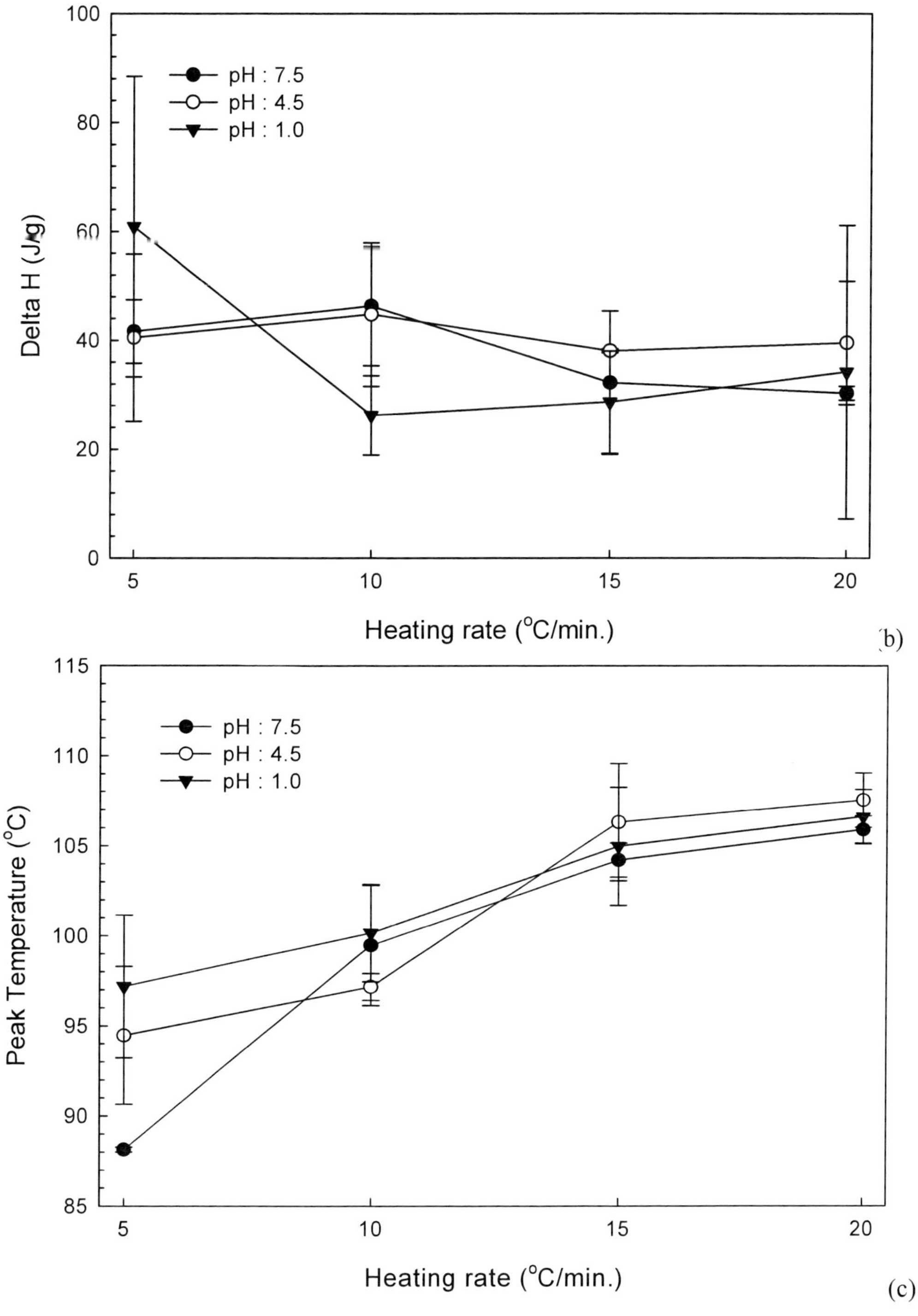

Figure 7. Thermal curing behavior of UF resins prepared under different pH conditions as a function of heating rates. (a) On-set temperature, (b) heat of reaction (ΔH), and (c) peak temperature.
*Reproduced from ref. [60] with by permission from John Wiley & Sons (© John Wiley & Sons, 2006).

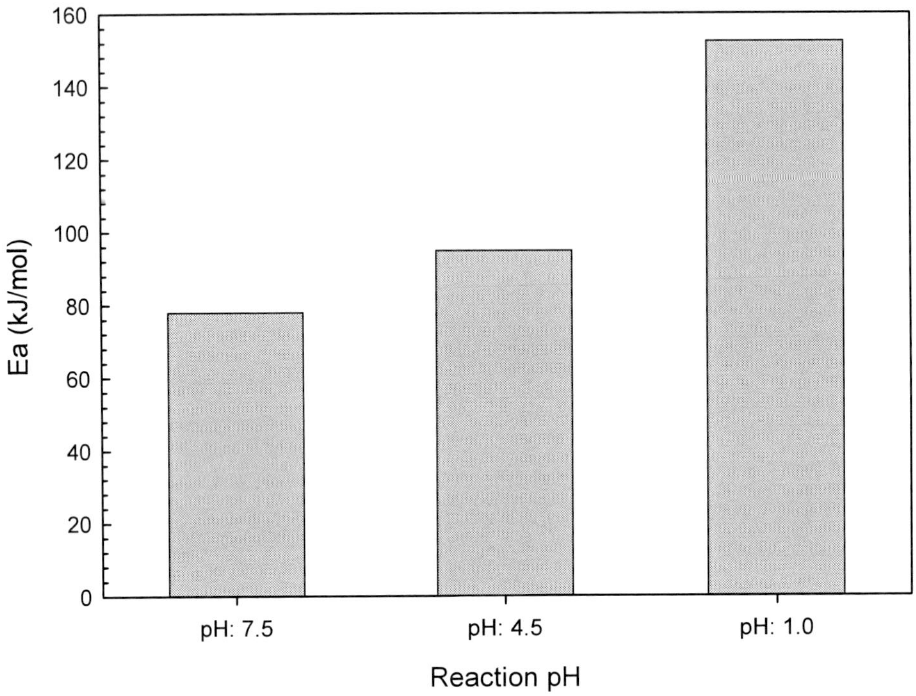

Figure 8. Activation energies of UF resins depending on synthesized pH conditions * Reproduced from ref. [60] with by permission from John Wiley & Sons (© John Wiley & Sons, 2006).

As shown in Figure 7 (a), the onset temperature generally increases with increasing heating rates. The highest onset temperature regardless of heating rates was found at the UF resin synthesized under weak acid condition, followed by the one prepared under the alkaline condition, and then the one prepared under the strong acid condition. These results indicate that the UF resin synthesized under strong acid condition provides lower onset temperature, which makes cure fast in the early stage of UF resin cure.

Figure 7 (b) demonstrates the change of reaction heat (ΔH) of UF resins during their curing. The ΔH slightly decreased with increasing heating rates for those UF resins synthesized under alkaline and weak acid conditions. But, differences between peak temperatures of the UF resin prepared under strong acid condition were small for different heating rates if the standard deviation was taken into account. In other words, the ΔH of the strong acid UF resin was the lowest compared with those of the others. Even though the onset temperature is an indicator of UF resin cure, or reactivity, the peak temperature is an important parameter of comparing the reactivity of UF resin [36]. The peak temperature is a temperature where the rate of cure reaches the maximum during a dynamic scan of the reaction. In general, the peak temperatures of UF resins increased with increasing heating temperatures (Figure 6). This result might be due to thermal lag that occurred in transferring heat from the capsule to the sample resin inside the capsule. The peak temperatures of UF resins were quite close for all heating rates.

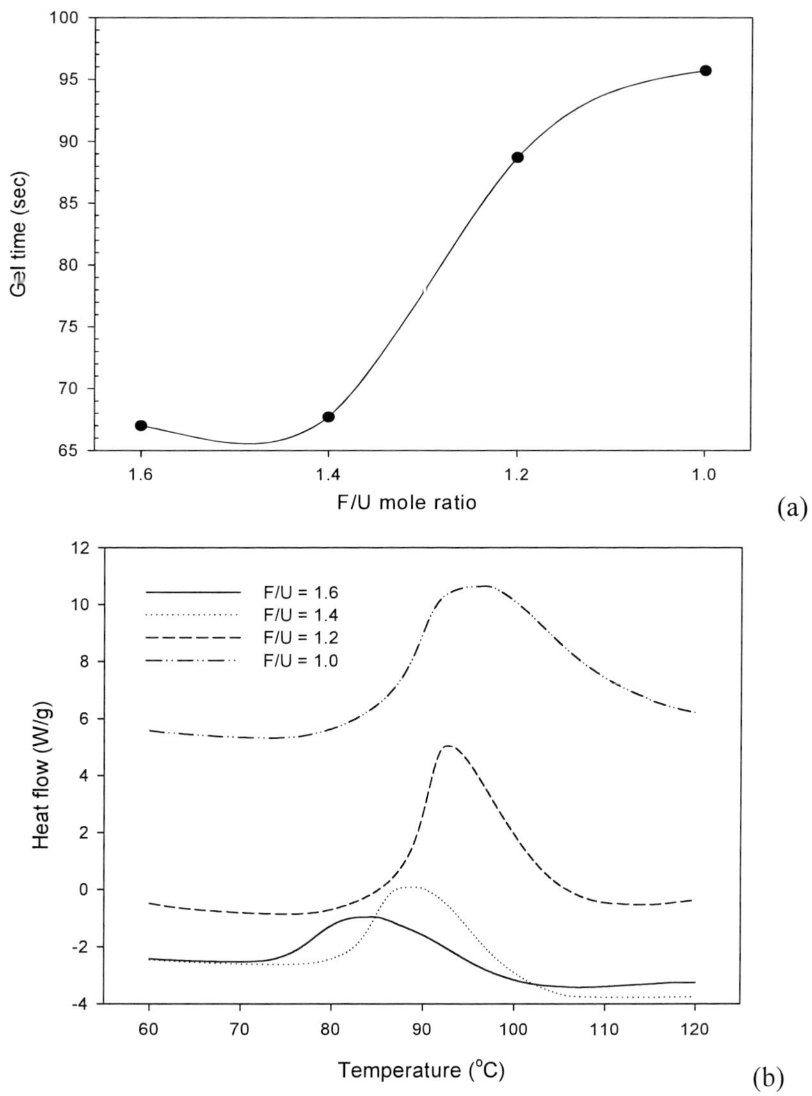

Figure 9. Thermal curing of UF resins at different F/U mole ratios. (a) Gel time, (b) DSC curves at 10°C/min.

* Reproduced from ref. [40] with by permission from John Wiley & Sons (© John Wiley & Sons, 2006).

The peak temperatures of UF resins at different heating rates were used to build a linear relationship between $ln(\beta/T_p^2)$ and $1/T_p$ as defined by the Eq. (1).

$$-\ln\left(\frac{\beta}{T_p^2}\right) = \frac{E}{RT_p} - \ln\left(\frac{ZR}{E}\right) \qquad (1)$$

where β is heating rate (°C/min.) and T_p is the peak temperature. The above equation provides a straight line between $-\ln(\beta/T_p^2)$ *and* $1/T_p$. *The* slope of the linear regression line was used to calculate the Ea. As shown in Figure 8, the Ea values of UF resins prepared under the alkaline, weak acid, and strong acid conditions were about 78 kJ/mol, 94.8 kJ/mol and 152.2 kJ/mol, respectively. These Ea values were greater than the reported one [37]. In general, the Ea increased as the pH levels decreased from alkaline to strong acid condition. This result suggests that the UF resin synthesized under strong acid condition requires more energy to start its curing process than those of the other UF resins prepared under weak acid and alkaline conditions.

The thermal curing results of the UF resins prepared at different F/U mole ratio are shown in Figure 9. The gel time of UF resins increased with decreasing the F/U mole ratio (Figure 9, a). In particular, the gel time rapidly increased when the F/U mole ratio decreased from 1.4 to 1.2. These results indicate that the curing reactivity of UF resin decreased with decreasing F/U mole ratio. This might be explained by a decrease in the availability of formaldehyde at lower F/U mole ratio. The result is quite compatible with those of onset and peak temperatures as shown in Figure 10, (a). Figure 9 (b) presents typical DSC curves of UF resins at different F/U mole ratios. All DSC curves obtained at the 10 °C/min show an exothermic peak at different temperatures. This exothermic peak could be attributed to the heat released from the polycondensation reaction of primary amino groups of un-reacted urea with methylolgroups (-CH_2OH) [38]. As the F/U mole ratio decreased, the exothermic peak temperature of UF resins increased, suggesting a decrease in the resin curing reactivity. And levels of heat flow of UF resin were much greater for lower F/U mole ratio. This was quite consistent with the result of heat of reaction (i.e. ΔH) as shown in Figure 10, (c).

Figure 10 illustrates thermal curing behaviors of UF resins with different F/U mole ratios. Both the onset and peak temperatures of UF resins at different F/U mole ratios are shown in Figure 10, (a). The onset temperature is defined as a temperature where the polymerization of UF resin starts under an acid condition (Figure 6). The onset temperature increased with decreasing the F/U mole ratio, which indicated a decrease in the resin reactivity at lower F/U mole ratio. By contrast, the peak temperature is defined as a temperature where the polymerization of UF resin reaches the maximum conversion rate. The peak temperature increased as the F/U mole ratio decreased. This result also suggested that the reactivity of curing reaction of UF resin was declined as the F/U mole ratio decreased.

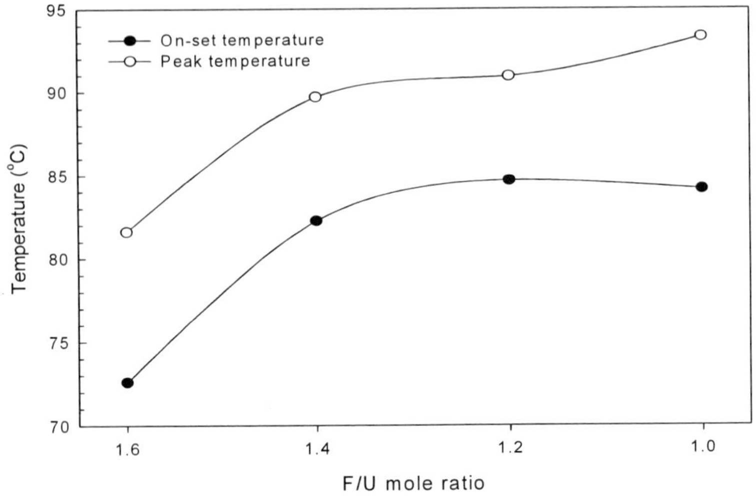

(a)

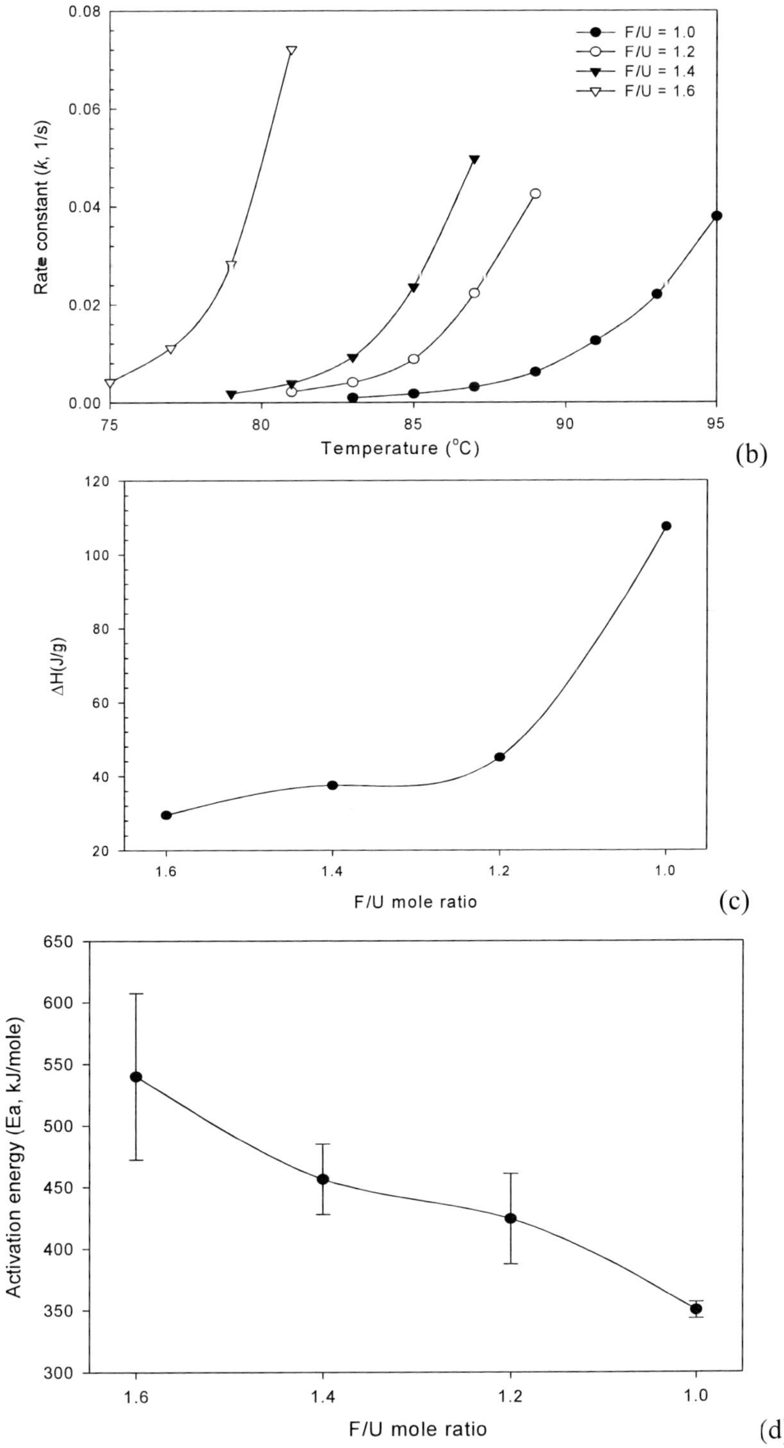

Figure 10. Thermal curing behavior of UF resins at different F/U mole ratios. (a) On-set and peak temperatures, (b) rate constant, (c) heat of reaction (ΔH), and (d) activation energy (E_a).
* Reproduced from ref. [40] with by permission from John Wiley & Sons (© John Wiley & Sons, 2006).

The gel time, onset and peak temperatures showed a similar trend in the resin reactivity, i.e., the curing reactivity decreased as the F/U mole ratio decreased. In addition, the rate constant (*k*) of curing reaction was compared for UF resins with different F/U mole ratios (Figure 10, b). In general, the rate constant of the curing reaction of UF resin increases as the temperature increases. As expected, the rate constant decreased as the F/U mole ratio decreased. This result was quite compatible with those of the gel time and peak temperature. In other words, a decrease of the rate constant provided a justification of decreases in the gel time and peak temperature, which were regarded as indicators of the curing reactivity of UF resin. Figure 10, (c) exhibits the heat of curing reaction (ΔH) of UF resin in the presence of acid hardener. The ΔH is defined the area under an exothermic thermogram of a DSC curve. The ΔH increased with decreasing the F/U mole ratio. This result was quite reasonable. Since the gel time and peak temperature increased with decreasing the F/U mole ratio, the area under the DSC curve was getting larger with decreasing the F/U mole ratio, which resulted in an increase of the ΔH. These results suggested that more energy was required to complete the cure of UF resin when a lower F/U mole ratio was used.

The Ea based on the Eq. (1) of UF resins with different F/U mole ratios is shown in Figure 10, (d). The Ea decreased with decreasing the F/U mole ratio. This result indicated that UF resin with lower F/U mole ratio requires less energy to spontaneously start the curing reaction than that with higher F/U mole ratio. In general, more branched polymers require greater Ea than those of less branched ones. So, it is believed that UF resin prepared at higher F/U mole ratio has more branched network polymer than that synthesized at lower F/U mole ratio. However, the composition of chemical species such as monomethylolurea, dimethylolurea, and trimethylolurea could provide different levels of branched polymer network after the cure of UF resin. Thus, further research work needs to employ ^{13}C-NMR spectroscopy to compare the composition of these chemical species depending on different F/U mole ratios.

3.3. Dynamic Mechanical Properties of UF Resins

Figure 11 demonstrates DMA curves of UF resin adhesives with the F/U mole ratio of 1.0. Definitions of thermomechanical parameters were also given in Figure 11. The storage modulus, E′ decreased to a minimum (E'_{min}), and then increased to a maximum (E'_{max}) as the temperature increased. The difference of storage modulus between the E'_{min} and E'_{max} was defined as ΔE′. The temperature where the E′ reached minimum was defined as the gel temperature (T_{gel}) of UF resin adhesive. Peak temperatures of reaching the maximum storage modulus (E'_{max}) and loss modulus (E''_{max}) were defined as T_1 peak temperature and T_2 peak temperature, respectively. The rigidity represented as the E′ initially decreased to a minimum and then reached a maximum. The initial decrease of E′ could be due to the softening of UF resin adhesives as the temperature increased. After the E'_{min}, the E′ started to increase toward a maximum. This was possibly due to the gelation of UF resin adhesive, where an infinite molecular network began to be formed. Thus, this temperature was presented as the gel temperature. Similar definition of the gel time was reported for an isothermal scanning of melamine-modified UF resins [39].

An increase of the E′ after the minimum was possibly ascribed to the change of network structure of UF resin adhesive from gel state to glassy state where the number of cross-linking increased as it went through curing process. A decrease of the E′ after the maximum could be due to combined effects of many factors. One of the factors might be devitrification of UF resin after its glass transition temperature (T_g). Another factor would be hydrolytic or thermal degradation of UF resin as the temperature increased. Thus, as the resin adhesive went through gelation, the E′ continuously increased to a maximum where the resin became vitrified. Further increase in temperature resulted in a decrease of the E′, which could be resulted from devitrification of UF resin.

Figure 11 also shows the presence of two peaks of the E′. This observation could be explained by the curing process of UF resin. In other words, the first peak was occurred due to a vitrification after its gelation while the second peak might be due to another vitrification followed by devitrification. Further work is necessary to understand the presence of these two peaks. Loss modulus (E″) of UF resin adhesive with the F/U mole ratio of 1.0 followed similar change to the E′. Initial decrease of E″ could be due to the softening of UF resin as the rigidity did. The E″ started to increase after reaching the minimum. This result also reflected the gelation of UF resin, where the polymerization reaction started to form the network that resulted in efficient energy dissipation.

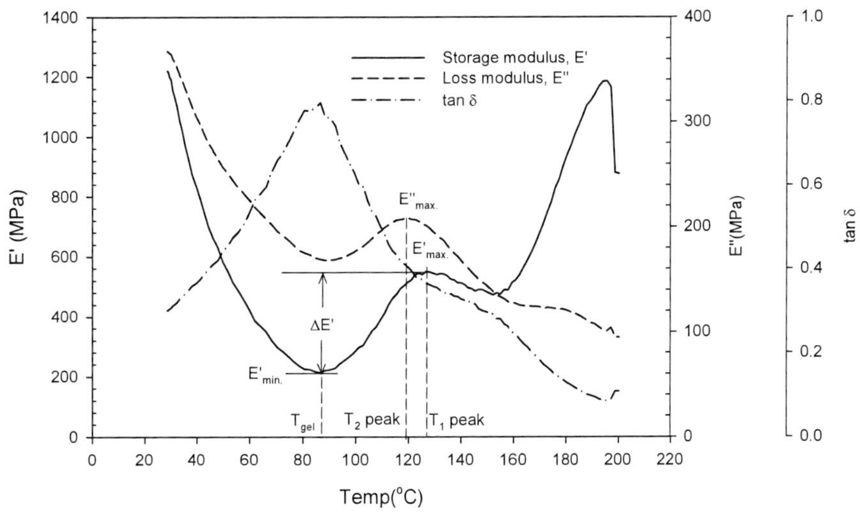

Figure 11. DMA curves of UF resins with F/U mole ratio of 1.0.
*Reproduced from ref. [65] with by permission from John Wiley & Sons (© John Wiley & Sons, 2008).

The E′ curves of UF resin adhesives with different F/U mole ratios are presented in Figure 12. All the E′ curves had similar pattern as the temperature increased. In other words, the E′ initially decreased to a minimum and then reached a maximum followed by a decrease. Figure 12 also showed the change of the E'_{max} of UF resin adhesive, depending on F/U mole ratio. As F/U mole ratio decreased, the E'_{max} increased to a maximum at the F/U mole ratio of 1.4 and then continuously decreased. This result indicated a reduction of the rigidity of UF resin as the F/U mole ratio decreased. In particular, lower E'_{max} of UF resin with lower F/U mole ratio of 1.0 could have provided the resin adhesive with lower cohesive adhesion strength. This result partially explains a deterioration of internal bond strength of particleboard bonded with UF resins of lower F/U mole ratio [13, 40].

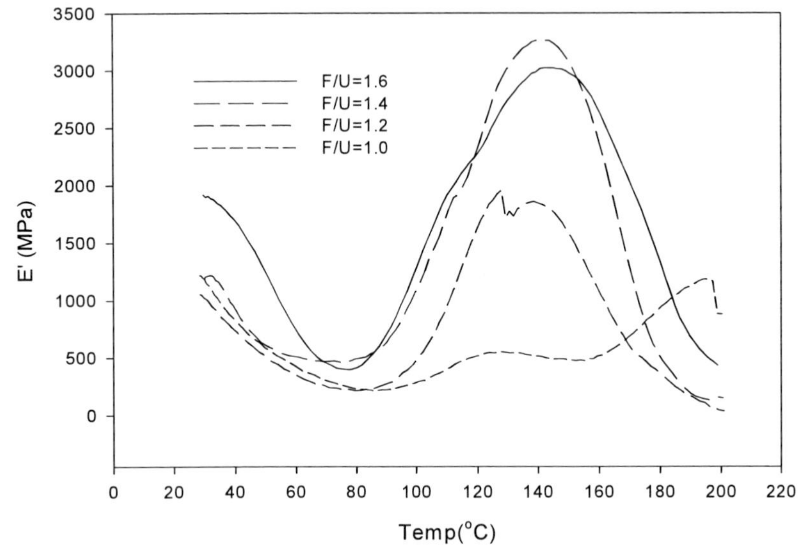

Figure 12. Typical storage modulus curves of UF resins with different F/U mole ratios.
*Reproduced from ref. [65] with by permission from John Wiley & Sons (© John Wiley & Sons, 2008).

However, the E'_{min} slightly decreased with decreasing F/U mole ratio, and showed not much difference. The $\Delta E'$, the difference between E'_{min} and E'_{max} also increased up to the F/U mole ratio of 1.4 and then decreased. This result indicated that the $\Delta E'$ was mainly dependent on the E'_{max} rather than the E'_{min}. In other words, the influence of F/U mole ratio to the rigidity was more predominant on the E'_{max} rather than the E'_{min}. Therefore, the $\Delta E'$ could be used as an indicator of the rigidity of UF resins. In fact, the $\Delta E'$ was used as a stiffening coefficient for the comparison of thermomechanical behaviors of different of adhesive systems [41].

Figure 13 (a) shows the E''_{max} of UF resin adhesives depending on F/U mole ratios. The E''_{max} showed similar pattern to that of the E'_{max} with decreasing F/U mole ratio. The loss modulus contributes the energy dissipation due to molecular friction owing to the viscose flow of a material. Thus, the result indicated that molecular friction of cured UF resin adhesive was reduced as F/U mole ratio decreased. Similar result was reported for PF resole resin [42]. This could be attributed to more branched network structure of UF resin with higher F/U mole ratio than those of lower F/U mole ratios. In other words, it seemed that UF resin of lower F/U mole ratio was more flexible than those of higher F/U mole ratios, which required less energy dissipation under the oscillation. In fact, it was reported that UF resins with the F/U mole ratio of 1.0 was predominantly composed of linear methylene linkages [30]. The gelation temperature (T_{gel}) and peak temperature of tan δ of UF resin adhesives depending on F/U mole ratio is shown in Figure 13 (b). In general, the gel temperature of UF resin adhesive increased with decreasing F/U mole ratio, but they were much different at the

F/U mole ratio below 1.4. This result indicated that higher F/U mole ratio resin reached faster gelling than that of lower F/U mole resin, showing a greater reactivity of UF resin of higher F/U mole ratio. This result is quite compatible with the results PF resin adhesives [42], and is supported by the results of DSC [40]. The peak temperature of tan δ of UF resin also showed a similar pattern to the gel temperature. This could be due to the gelation of UF resin adhesive as the UF resin adhesive started to increase the number of cross-linking.

Figure 13 (c) shows peak temperatures of both the E′ and E″ of DMA curves. T_1 peak and T_2 peak temperatures is the peak temperature of the E'_{max} and E''_{max}, respectively. As the F/U mole ratio decreased, the T_1 peak temperature slightly increased up to the F/U mole ratio of 1.4, and then gradually decreased afterward. This result suggests that the UF resin with F/U mole ratio of 1.6 reaches a maximum rigidity faster than the other UF resins, which resulted in a decreased T_1 peak temperature of the E′. In other words, a decreased T_1 peak temperature with decreasing F/U mole ratio could be due to smaller value of the E'_{max} of UF resin with lower F/U mole ratio. By contrast, the T_2 peak temperature gradually increased as the F/U mole ratio decreased. This result indicated that as the F/U mole ratio decreased, it took more time for UF resins to reach a point of the maximum of energy dissipation. The maximum tan δ of UF resin adhesives depending on F/U mole ratios is shown in Figure 13 (d). As the F/U mole ratio decreased, the maximum tan δ proportionately increased with F/U mole ratio. The gelling of UF resin resulted in a minimum E′ and a maximum tan δ around the same temperature. In general, the maximum tan δ results from the gelling or vitrification of thermosetting resin. Thus, this result could be due to increasing gel temperatures as F/U mole ratio decreased. Furthermore, an increase of the maximum tan δ with decreasing F/U mole ratio suggested that an elastic component of UF resin decreased while a viscose component increased under oscillation. In other words, damping behavior of UF resin increased with decreasing F/U mole ratio. A greater amount of the energy used to deform was dissipated into heat in UF resin of lower F/U mole ratio than the one of high F/U mole ratio.

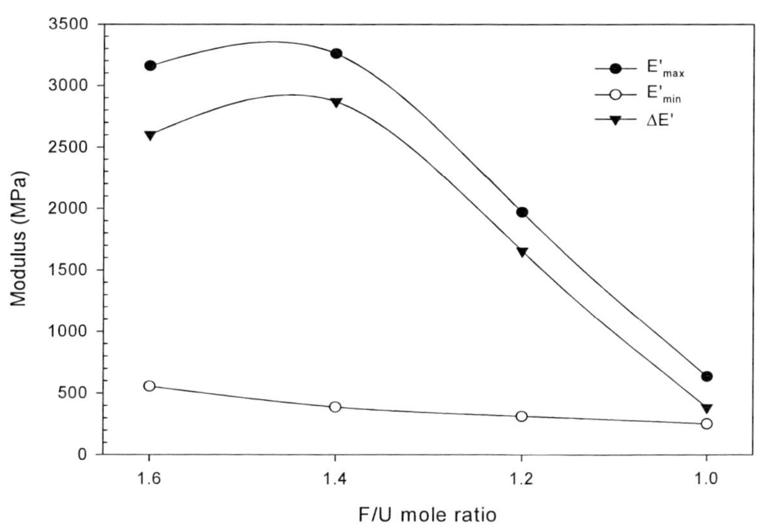

(a)

Figure 13. (Continued).

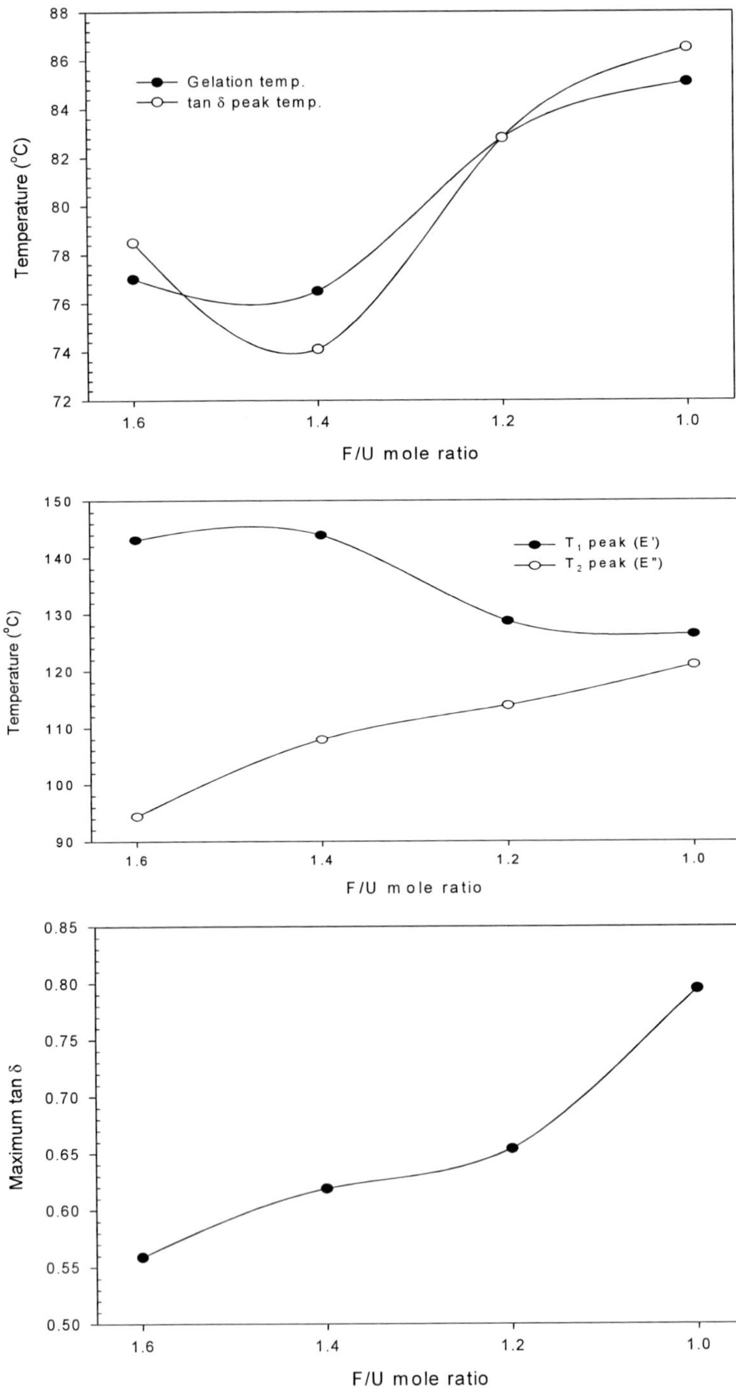

(b)

(c)

(d)

Figure 13. Thermomechanical curing behavior of UF reins with different F/U mole ratios. (a) E'_{max}, E'_{min}, and $\Delta E'$, (b) tan δ, (c) peak temperatures of E′ and E″, and (d) maximum tan δ values. *Reproduced from ref. [65] with by permission from John Wiley & Sons (© John Wiley & Sons, 2008).

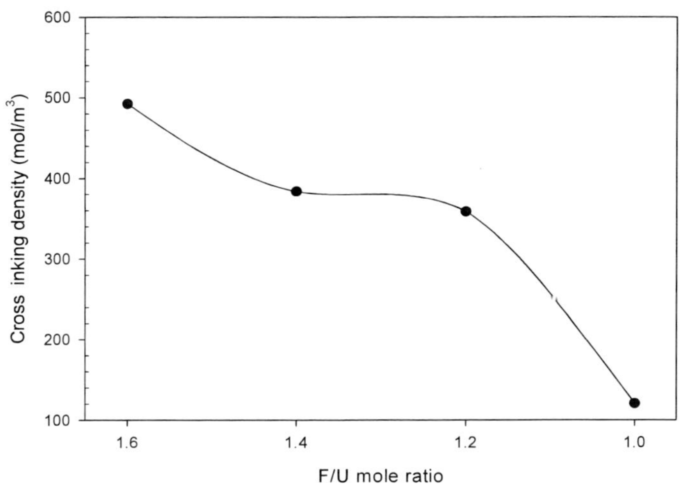

Figure 14. The calculated cross-linking density of UF resins with different F/U mole ratios. *Reproduced from ref. [65] with by permission from John Wiley & Sons (© John Wiley & Sons, 2008).

The kinetic rubber theory of elasticity makes it possible to calculate the experimental cross-linking density based on the storage modulus of DMA using the Eq. (2) shown below [43-44].

$$E' = 3\rho_c RT \tag{2}$$

where the E' is the storage modulus at $T_g + 40°C$ in the rubber plateau, ρ_c is the cross-linking density, R is the gas constant, and T is the absolute temperature (K). The gel temperature of UF resins was assumed as a glass transition temperature (T_g) for the calculation of the cross-linking density of UF resins. The calculated results were shown in Figure 14. As expected, the cross-linking density of UF resin decreased with decreasing F/U mole ratio. This result provided a theoretical background of explaining lower rigidity of UF resin with lower F/U mole ratio. Furthermore, it was believed that this would have also contributed to a poor adhesion performance of UF resin with lower F/U mole ratio when they were applied for particleboard manufacture.

3.4. Hydrolytic Stability of UF Resins with Different F/U Mole Ratios

Figure 15 shows the hydrolytic stability of cured UF resins prepared at different F/U mole ratios. As the F/U mole ratio decreases, the mass losses of two different cured UF resins with different particle sizes (180 μm and 250 μm) increased and then decreased (Figure 15, a). Smaller particle sizes influenced hydrolytic stability of cured UF resins. This could be due to larger surface areas of the smaller particles than those of the larger particles. A similar trend was also found for the liberated formaldehyde concentration after acid hydrolysis (Figure 15, b). It is believed that lower mass loss at the F/U mole ratio of 1.6 could be due to a greater cross-linking density. A decrease in the mass loss and liberated formaldehyde concentration of cured UF resin with an F/U mole ratio of 1.4 could be due to a high branched network structure of the resin. But, in general, these results indicate that hydrolytic stability of cured UF resin improved as the F/U mole ratio decreases. In other words, UF resin of lower F/U mole ratio is more resistance to hydrolysis than those of higher F/U mole ratios. These results might be related to the molecular structure of cured UF resin. It is known that UF resin of higher F/U mole ratio is much more branched than those of low F/U mole ratio [18, 25-26, 45]. A greater degree of branch of UF resin has a greater probability of exposing the methylol groups to hydrolysis, which subsequently increases the mass loss and the concentration of liberated formaldehyde [25]. However, it was reported that UF resins of low F/U mole ratio were less branched and more linear in structure [18]. Thus, the linear structure of low mole ratio UF resins has fewer number of methylol groups exposed to hydrolysis, which will consequently improve hydrolytic resistance.

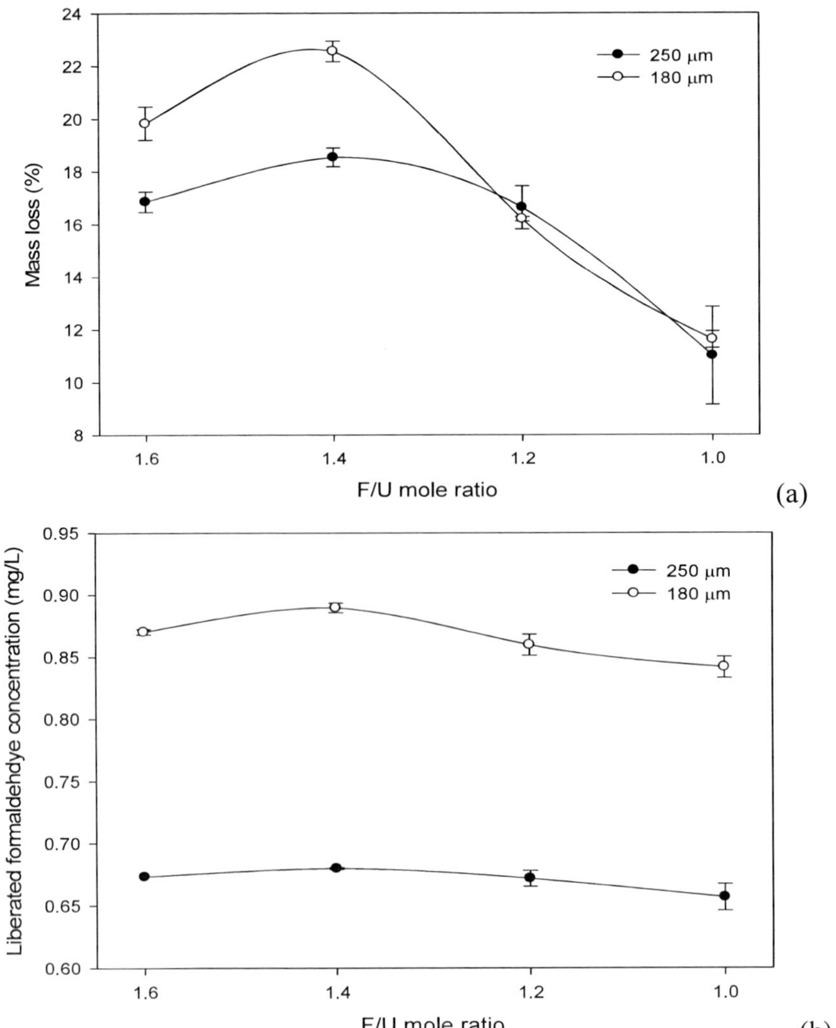

Figure 15. Hydrolytic stability of cured UF resins with different F/U mole ratios. (a) Mass loss, and (b) liberated formaldehyde concentration.

*Reproduced from ref. [59] with by permission from Springer (© Springer, 2011).

3.5. Adhesion Performance of UF Resins

Figure 16 explains shear adhesion strengths of the resins prepared under three different reaction pH conditions, and cured with three different hardener types. For the resins prepared under alkaline-acid condition, the adhesion strength was quite close to each other when hardened with ammonium chloride and zinc nitrate, but the poorest adhesion strength under the alkali-acid condition was found when the resin was hardened with ammonium citrate. For weak-acid and strong acid reaction condition, the adhesion strength was in order of ammonium chloride, ammonium citrate, and zinc nitrate. In other words, the strongest adhesion strength was found when the resin prepared under strong acid condition was hardened with ammonium chloride. This result was quite interesting in terms of the chemical species present in the resin prepared under strong acid reaction condition. Soulard *et al.* [17] reported that internal bond strength of particleboard increased to a maximum and then decreased as the amount of uron in the resin increased from 0% to 75%. Thus, it was believed that the strongest adhesion strength of the resin prepared under the strong acid condition is responsible for the presence of uronic structures in the resin.

FEs of particleboards bonded with UF resins were shown in Table 8. Particleboards bonded with UF resins prepared under strong acid condition showed the least FEs followed by alkaline, and weak acid condition. This result is related to the amount of free formaldehyde present in the resin. Thus, strong acid condition for UF resin synthesis would be an option to reduce FE of UF resin. However, the IB strengths of particleboard showed quite contrasting results: particleboard prepared with UF resin reacted in strong acid condition showed the poorest strength followed by weak acid and alkaline conditions. This result also suggests that the UF resin synthesized under strong acid condition needs to be modified to improve its adhesive bond strength in particleboard.

Figure 17 shows the properties of PB bonded with UF resins at different F/U mole ratios. As shown, the amount of free formaldehyde present in the UF resin decreased with decreasing the F/U mole ratio. This result also makes it possible to explain changes of the gel time and onset temperature as a function of F/U mole ratios. As the UF resin cures under the acidic condition produced by the added hardener (i.e. NH_4Cl), a decrease of the free formaldehyde amount with decreasing F/U mole ratio could retard to begin the cure of UF resin, which resulted in a longer gel time and higher onset temperature. In addition, the amount of free formaldehyde in UF resin was closely related with the FE (FE) of particleboard (PB).

In general, the FE of PB was heavily dependent on the amount of free formaldehyde present in UF resin before it was cured. This result was quite compatible with other findings [13]. This result also suggested that the amount of free formaldehyde in UF resin made a great contribution for the FE of PB. The FE of PB bonded with UF resins prepared at different F/U mole ratios rapidly decreased up to the F/U mole ratio of 1.2, and then slightly decreased as the F/U mole ratio decreased to 1.0 (Figure 12). This result showed that the F/U mole ratio below 1.2 was more effective in reducing the FE of PB. Thus, the F/U mole ratio should be below 1.2, preferably 1.0, in order for UF resin to lower the FE of PB. This result was in a good agreement of other result [9].

Table 8. Properties of the UF resins prepared at different F/U mole ratios

F/U mole ratio	Non-volatile solid content (%)	Viscosity (cps)	Specific gravity
1.6	50.1	240	1.26
1.4	51.7	248	1.25
1.2	53.3	254	1.25
1.0	54.5	160	1.21

* Reproduced from ref. [40] with by permission from John Wiley & Sons (© John Wiley & Sons, 2006).

Table 9. Mechanical properties of PB bonded with UF resin at different F/U molar ratios

F/U ratio	Board MC (%)	Board density (kg/m^3)	MOR (kgf/cm^2)	MOE ($x10^3$ kgf/cm^2)	IB strength (kgf/cm^2)
1.0	5.7	694	136.3	23.5	8.8
1.2	5.9	746	185.5	27.6	9.7
1.4	6.9	664	143.2	20.8	9.1
1.6	6.8	684	138.7	25.8	9.9

* Reproduced from ref. [40] with by permission from John Wiley & Sons (© John Wiley & Sons, 2006).

The properties of PB bonded with UF resins of different F/U mole ratios were also presented in Table 9 and Figure 18. The moisture content (MC) of PB ranged from about 6% to 7% for all F/U mole ratios. The density of PB was around the target density of 700 kg/m^3 with the exception of PB bonded with the UF resin prepared at the F/U mole ratio of 1.2. Both modulus of rupture (MOR) and modulus of elasticity (MOE) were not changed much for all F/U mole ratios. The IB strengths of PBs bonded with UF resins prepared at the different F/U mole ratios were also shown in Table 9. The IB strength gradually decreased with decreasing the F/U mole ratio. This result could be attributed to a decreased reactivity of UF resin with a lower F/U mole ratio as shown by the gel time, peak temperature, and rate constant. But, the variations of IB strength were relatively large compared to other properties.

Figure 18 also illustrates both thickness swelling and water absorption of PB bonded with UF resins at different F/U mole ratios. As expected, the thickness swelling of PB increased with decreasing F/U mole ratio of UF resin. This result was consistent with the IB strength. As the F/U mole ratio decreased, the curing reactivity of UF resin decreased, and consequently resulted in a lower IB strength. Lower IB strength allows more water molecules to penetrate into the board, resulting in a greater thickness swelling. In general, thickness swelling has a negative relationship with IB strength of PB [9]. Water absorption of PB showed different behavior with decreasing F/U mole ratio. But, water absorption increased for lower F/U mole ratio. These results indicated that lowering F/U mole ration of UF resin cause a loss of dimensional stability of PB at the expense of reduced FE.

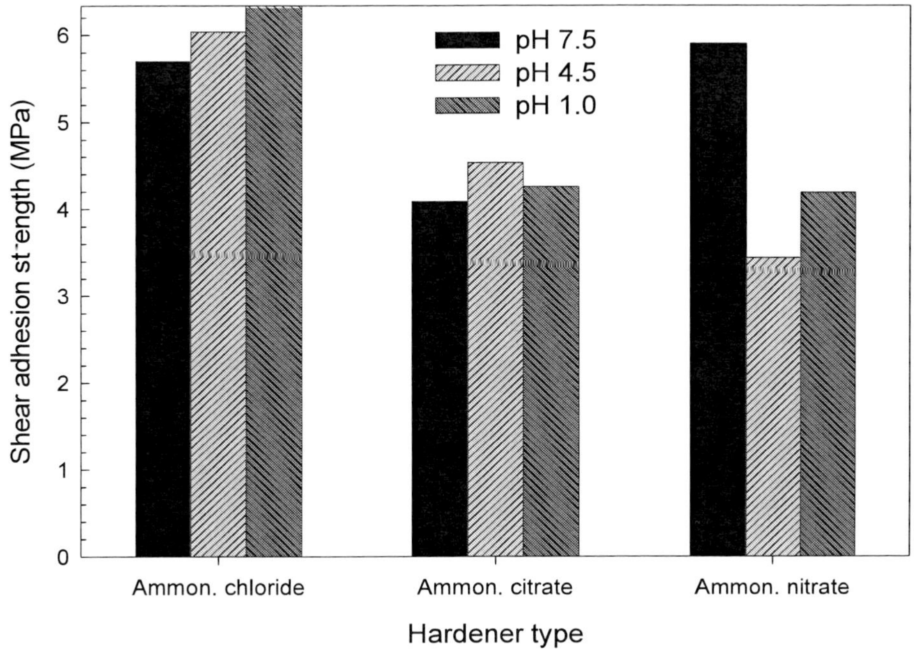

Figure 16. Shear adhesion strength of UF resins synthesized under three different pH conditions, and cured with different hardener types.
*Reproduced from ref. [36] with by permission from John Wiley & Sons (© John Wiley & Sons, 2003).

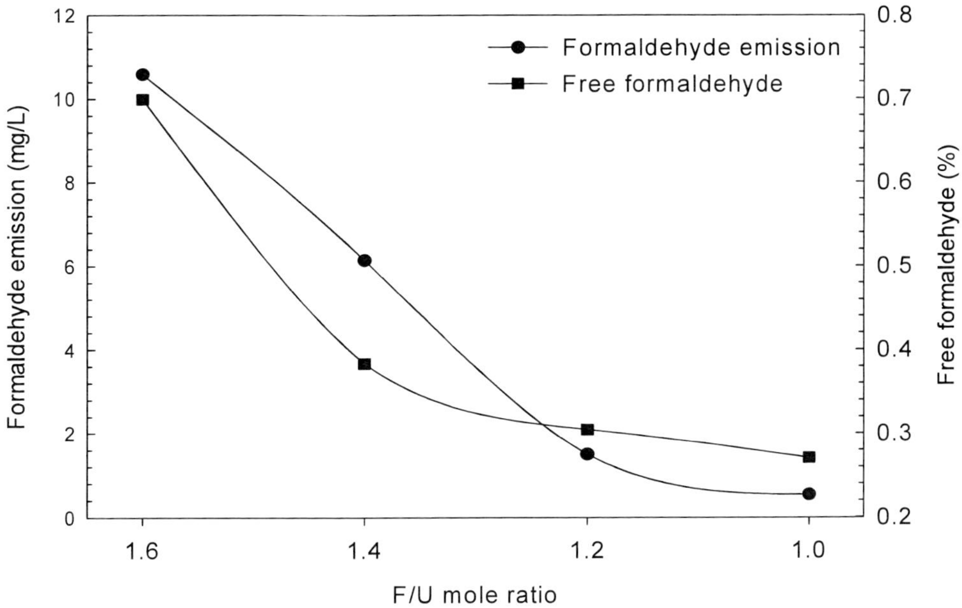

Figure 17. Formaldehyde emission and free formaldehyde of particleboard bonded with UF resins at different F/U mole ratios.
*Reproduced from ref. [40] with by permission from John Wiley & Sons (© John Wiley & Sons, 2006).

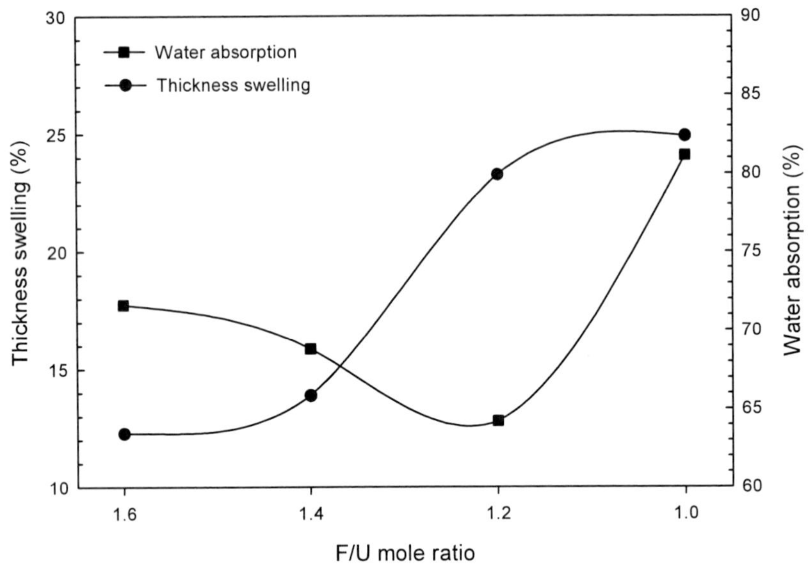

Figure 18. Thickness swelling and water absorption of particleboard bonded with UF resins at different F/U mole ratios.
*Reproduced from ref. [40] with by permission from John Wiley & Sons (© John Wiley & Sons, 2006).

4. Morphology, Microstructure and Crystalline Structure of Cured UF Resins

4.1. Morphology of Cured UF Resins

Figure 19 illustrates typical FE-SEM images of the exterior and fracture surfaces of cured UF resin films. The exterior surface of the cured UF resin was quite flat and smooth (Figure 19, a). However, the fracture surface showed linear marks, which were believed to be created by the strain of the sample shrinkage under liquid nitrogen (Figure 19, b). In addition, different sizes of pores were observed on the fracture surface, which might have resulted by the evaporation of water during its curing process.

As shown in Figure 20 (a), low F/U mole ratio UF resins illustrated numerous needle-like structures on the surface, which were believed to be a part of the crystalline structures. When it was etched by the acid, the surface displayed a typical spherical structure (Figure 20, e). The spherical structures of UF resins have been reported for F/U mole ratio lower than 1.2 [46-47]. Lower F/U mole ratio UF resins have colloid particles, which are coalesced into clusters during the aging process. These clusters are known to form spherical structures in cured UF resins [48]. The exterior surface of the cured UF resins with F/U mole ratios higher than 1.2 exhibited many crinkles. These crinkles were believed to have been formed by the shrinkage of the UF resin during its curing process (Figure 20, b & f). Similar images were obtained for all other samples, and the number of crinkles after the acid etching decreased as the F/U mole ratio increased (Figure 20, g & h). However, the cured UF resins with F/U mole ratios higher than 1.2 did not show any spherical structures after the acid etching.

(a)

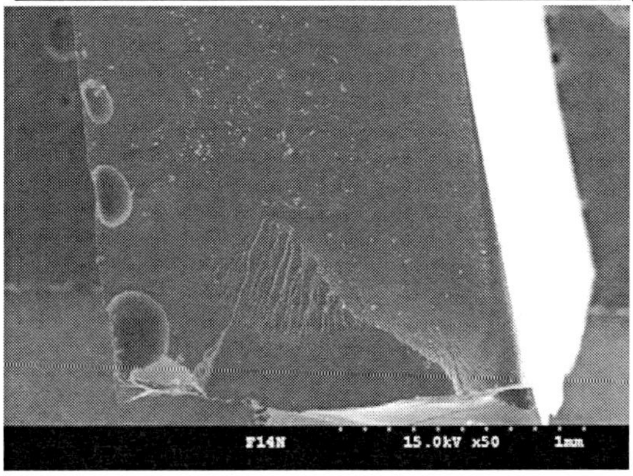

(b)

Figure 19. Typical morphologies of both exterior and fracture surfaces of cured UF resin (F/U = 1.4) films.
*Reproduced from ref. [70] with by permission from Mokchae Konghak (©Mokchae Konghak, 2011).

To compare the exterior surface's microstructure, FE-SEM was also applied to the fracture surface of the cured UF resin films. Figure 21 illustrates fracture surface images of cured UF resins before and after the acid etching. Unlike the exterior surface, the fracture surface of the cured UF resins with an F/U mole ratio of 1.0 demonstrated the presence of the spherical structures even before the acid etching (Figure 21, a). After the acid etching, the spherical structures were dominant on the surface. As mentioned earlier, the spherical structures usually occur in UF resins with F/U mole ratios lower than 1.2 [46-47]. In fact, it was reported that filament-like colloidal aggregates were initially formed in UF resin, and then eventually changed to super-clusters by coalescence during the aging process [49]. Thus, the observed structures for low F/U mole ratio UF resins are quite a normal phenomenon. After the acid etching as a simulation of the hydrolysis process, the fracture surface exhibited much more prominent spherical structures for the 1.0 F/U mole ratio cured UF resins (Figure 21, e). This result indicates that the acid etching apparently removed the inter-spherical regions or amorphous regions by the hydrolytic degradation process. This is believed to occur because the amorphous regions of the cured UF resin are easily hydrolyzed by the acid, so the remaining the spherical structures are much more resistant to the hydrolysis. In other words, these spherical structures could be arranged in a very ordered way that contributed to a crystalline part of the cured UF resins with low F/U mole ratios. Therefore, it is believed that these spherical structures deliver a greater resistance to the hydrolytic degradation of cured UF resins.

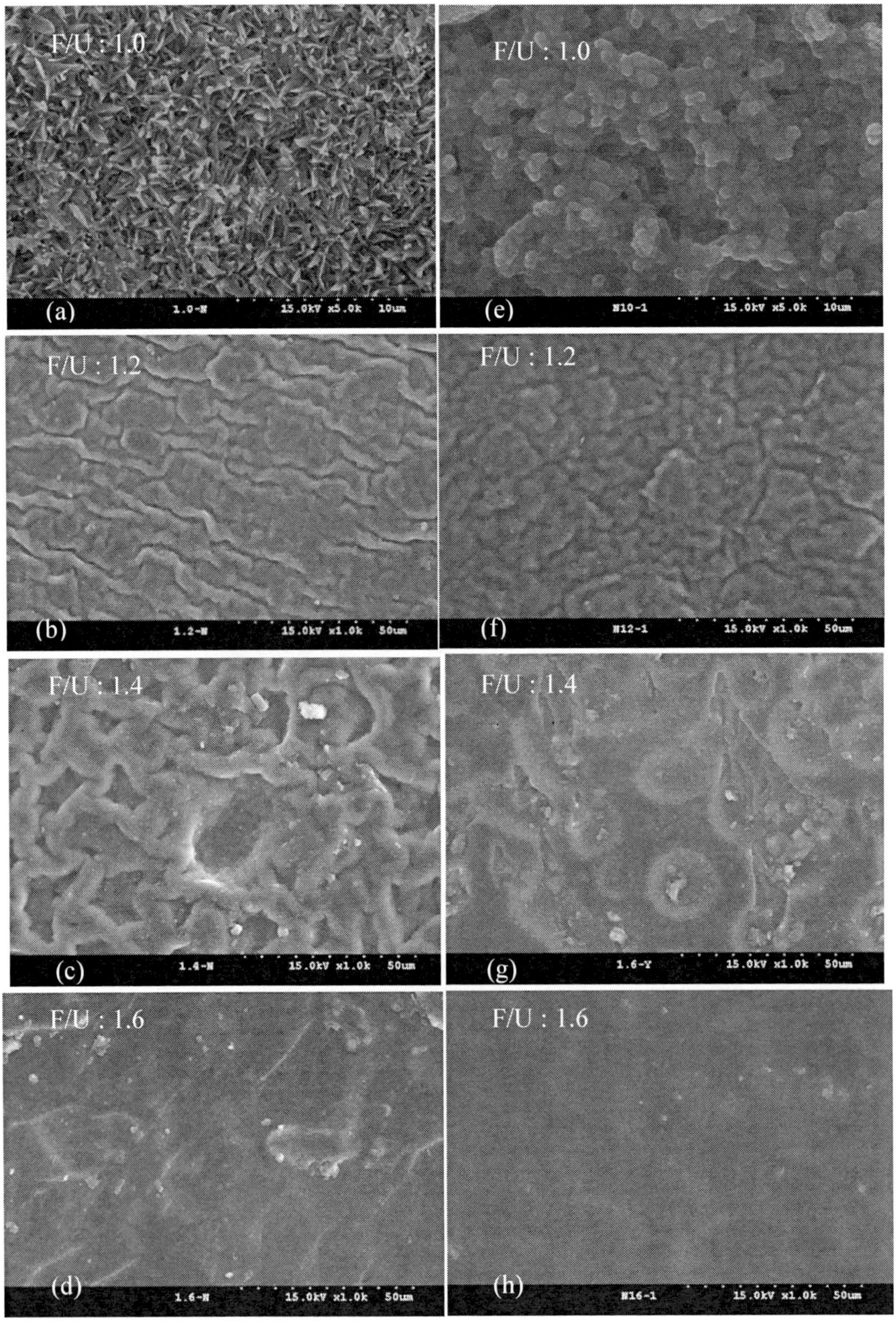

Figure 20. Typical exterior surfaces of cured UF resin with different F/U mole ratios. (a) ~ (d) No etching, and (e) ~ (h) with etching.

*Reproduced from ref. [70] with by permission from Mokchae Konghak (©Mokchae Konghak, 2011).

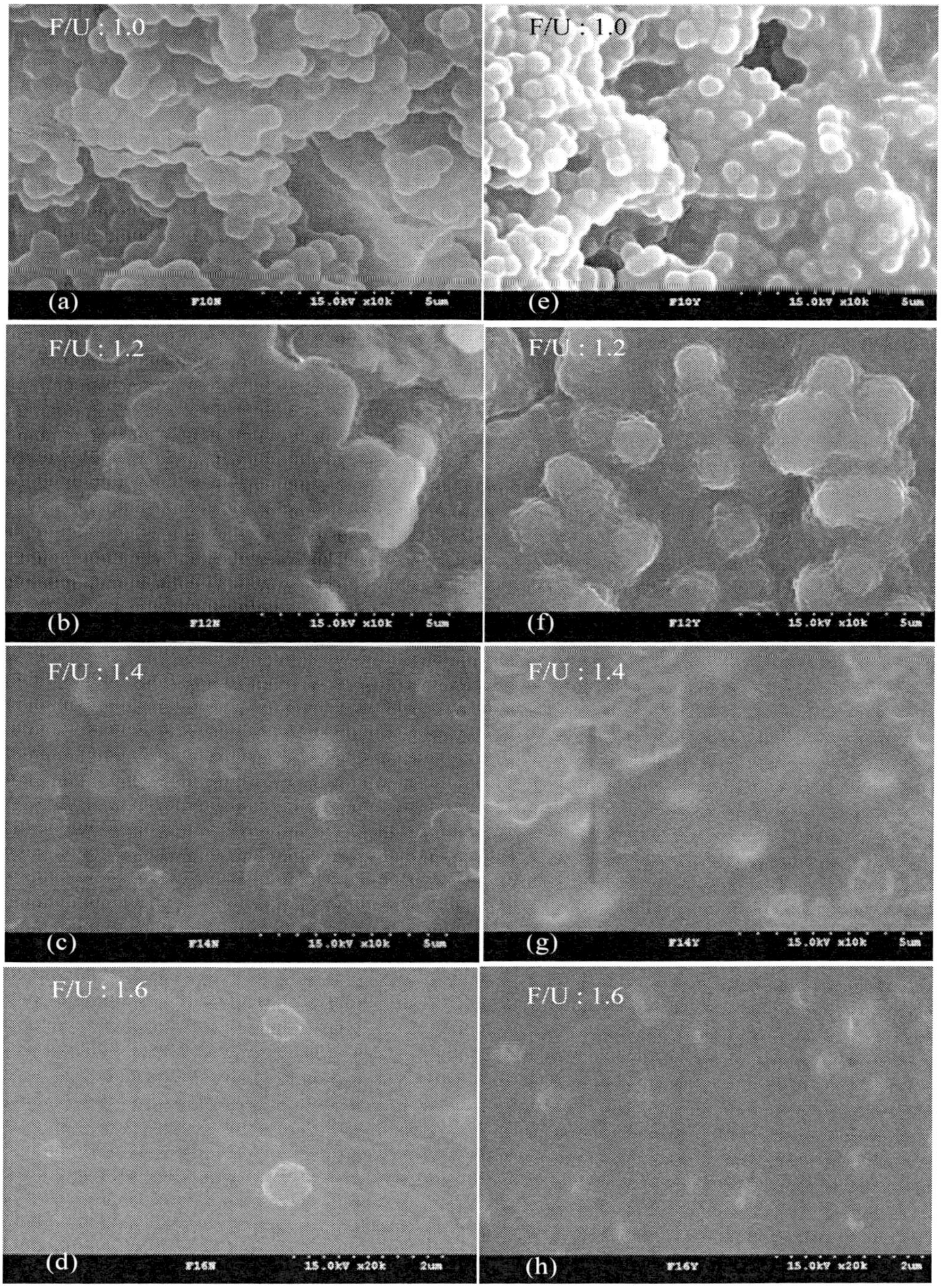

Figure 21. Typical fracture surfaces of cured UF resin with different F/U mole ratios. (a) ~ (d) no etching, and (e) ~ (h) with etching.
*Reproduced from ref. [70] with by permission from Mokchae Konghak (©Mokchae Konghak, 2011).

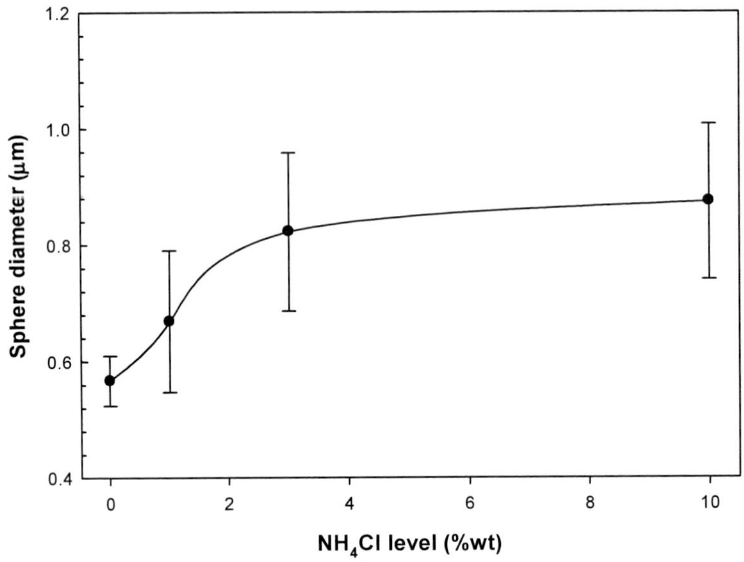

Figure 22. Sphere diameter of the cured UF resins of F/U mole ratio 1.0 at different NH_4Cl addition levels.
*Reproduced from ref. [58] with by permission from John Wiley & Sons (© John Wiley & Sons, 2011).

Similar results were also obtained for the cured UF resins with the F/U mole ratio of 1.2 (Figure 21, b & f). However, the number of the spherical structures was much lower than those of a 1.0 F/U mole ratio cured UF resins (Figure 21, a). The spherical structures on the fracture surface were quite definite before the acid etching (Figure 21, c & g). Although the frequency of their number of occurrence and size were fairly limited and small, the fracture surface still revealed spherical structures (Figure 21, d). Nevertheless, the acid etching of the fracture surface increased the number of spherical structure (Figure 21, h). However, the presence of the spherical structure for cured UF resins with F/U mole ratios higher than 1.4 have not been confirmed yet. This result indicates that spherical structures have been found with the cured UF resins of the higher F/U ratios of both 1.4 and 1.6 even though they are less dominant in terms of their number of occurrence and size.

Using FE-SEM images of the cured UF resins of F/U mole ratio of 1.0, an image analysis was done to measure the diameter of the spherical structure as a function of the NH_4Cl level. The results are shown in Figure 22. As the NH_4Cl level increased, the diameter increased up to 3% NH_4Cl and then levelled off. This might be due to an increase in the cross-linking density in the cured resin as the NH_4Cl level increased. In other words, the more NH_4Cl was added, the more acidic condition in curing UF resin, which provided a greater cross-linking density. As reported by Johns and Dunker [48], a colloidal structure of UF resin with low F/U mole ratio was formed by the presence of double layer of formaldehyde. Thus, an increased amount of NH_4Cl in UF resin could disrupts the double layer of formaldehyde, which subsequently increased the diameter of the spherical structure of the UF resin. But, the level-off of the diameter from 3% to10% NH_4Cl levels could be related to the amount of free formaldehyde in the UF resin. In other words, 10% NH_4Cl addition level did not influence to the spherical structure because 1.0 F/U mole ratio UF resin contained much smaller amount of the free formaldehyde than that of the counterpart. Thus, low free formaldehyde did not disrupt the double layer although high NH_4Cl was added in the resin.

4.2. Chemical Elements in Cured UF Resins

An energy dispersive spectroscopy (EDS) combined with the FE-SEM was used to understand the distribution of chemical elements in cured UF resins. Figure 23 shows a typical EDS result of cured UF resins of F/U mole ratio of 1.0 at different NH_4Cl levels. As expected, four chemical elements such as carbon (C), nitrogen (N), oxygen (O), and sodium (Na) were detected for the resins hardened without adding NH_4Cl. The N element resulted from the urea while both C and O elements were from either urea or formaldehyde used as raw materials. The Na element resulted from sodium hydroxide (NaOH) that has been used for the pH control during UF resin synthesis. The C distribution was the most abundant among four elements. This could be due to the fact that both urea and formaldehyde used as two major raw materials had carbon in common and cured UF resin also had methylolgroups (CH_2OH), methylene linkages ($-CH_2-$). The N element was obviously from the amine groups of urea, while the O element was carbonyl groups of both urea and formaldehyde. Both N and O elements were less abundant that that of the C element. As expected, the Na element was the least abundant among four elements.

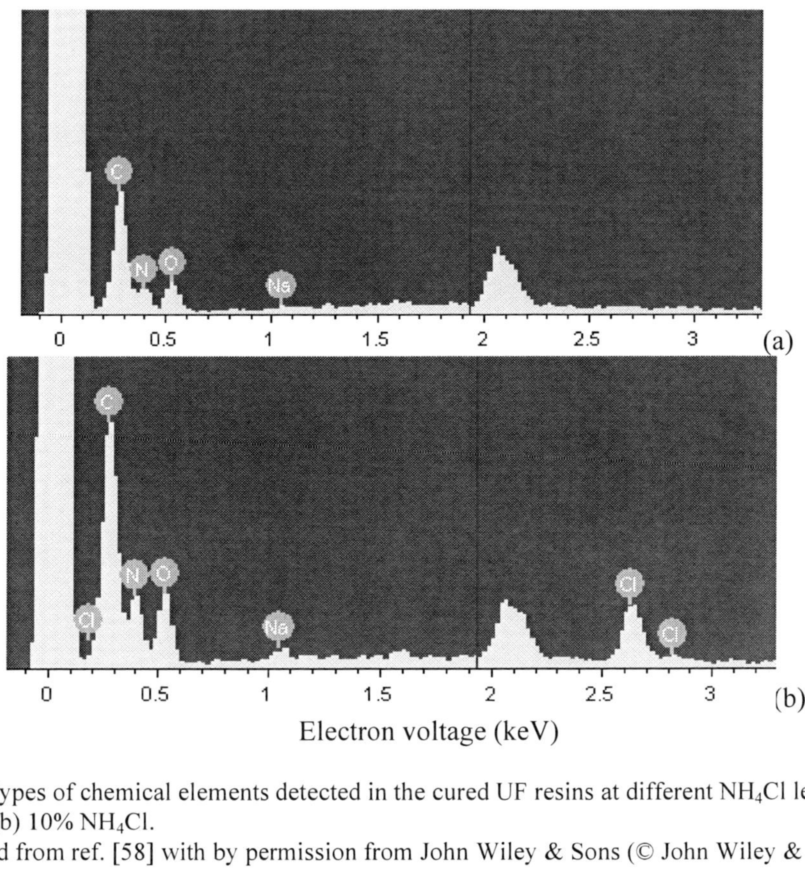

Figure 23. Types of chemical elements detected in the cured UF resins at different NH_4Cl levels. (a) 0% NH_4Cl and (b) 10% NH_4Cl.

*Reproduced from ref. [58] with by permission from John Wiley & Sons (© John Wiley & Sons, 2011).

The chloride (Cl) element was detected when the NH_4Cl was added. As expected, the peak intensity of Cl at 10% NH_4Cl was greater than those of either 1% or 3% NH_4Cl. Three peaks of Cl element observed at 0.25 keV, 2.65 keV and 2.8 keV by the EDS were resulted from the electron valence of K_α, K_β and L, respectively. This could be ascribed to different x-ray radiation energy from difference electron valences of the Cl [50]. The EDS was also employed to map the chemical elements at the facture surface, i.e. element distribution. The peak at 2.1 keV ascribed to the presence of platinum that had been used for coating the facture surface.

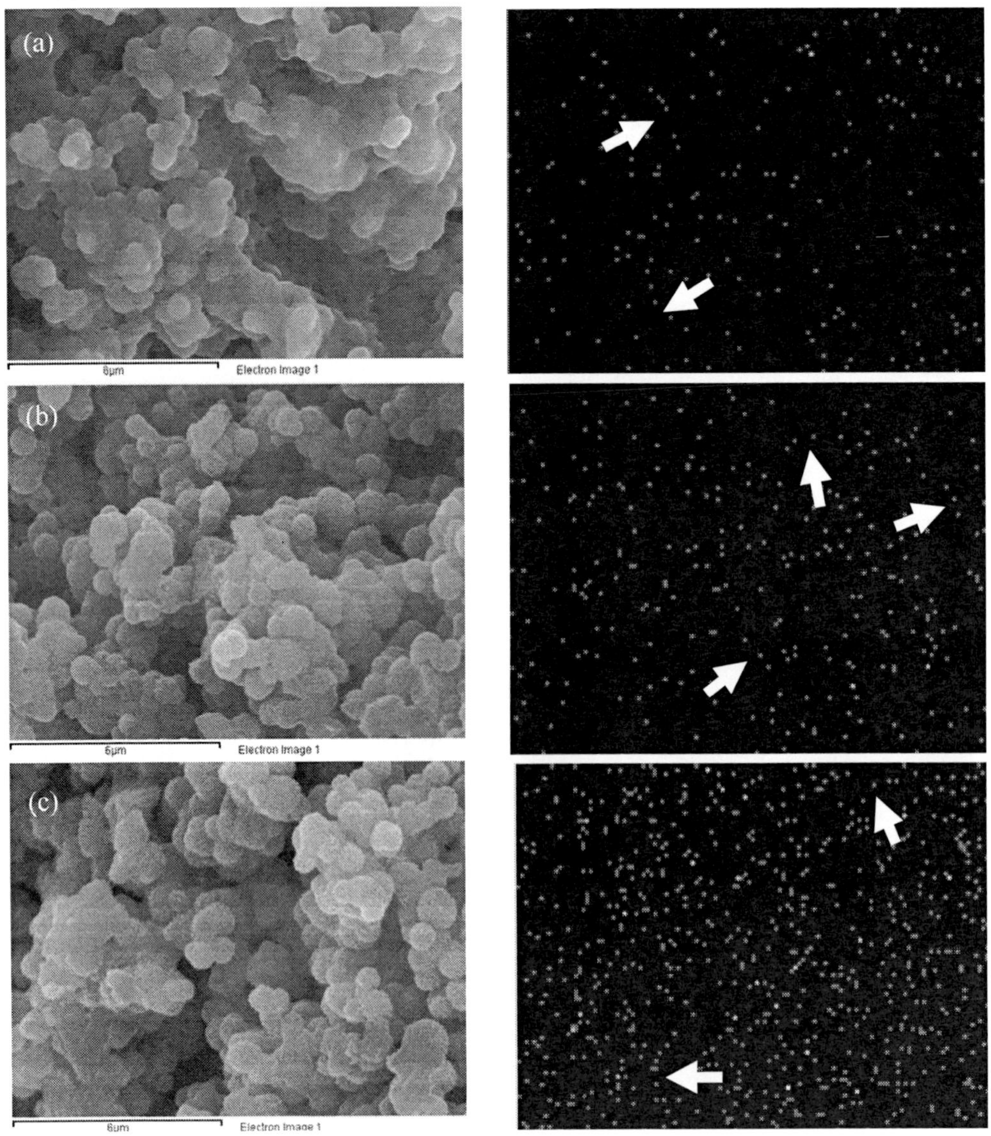

Figure 24. FE-SEM micrographs and corresponding Cl distributions of cured UF resins of the F/U mol ratio of 1.0 at different NH_4Cl levels. (a) 1% NH_4Cl, (b) 3% NH_4Cl, and (c) 10% NH_4Cl.
*Reproduced from ref. [58] with by permission from John Wiley & Sons (© John Wiley & Sons, 2011).

In general, UF resin is polymerized under acidic condition through a curing process. Usually, a hardener or catalyst is added into the UF resin in order to obtain an acidic condition. For example, the NH_4Cl as a hardener reacts with free formaldehyde in UF resin to form hydrochloric acid in the cured state [51]. The reaction is given as the Eq. (3):

$$6CH_2O + 4NH_4Cl \rightarrow (CH_2)_6N_4 + 6H_2O + 4HCl \quad (3)$$

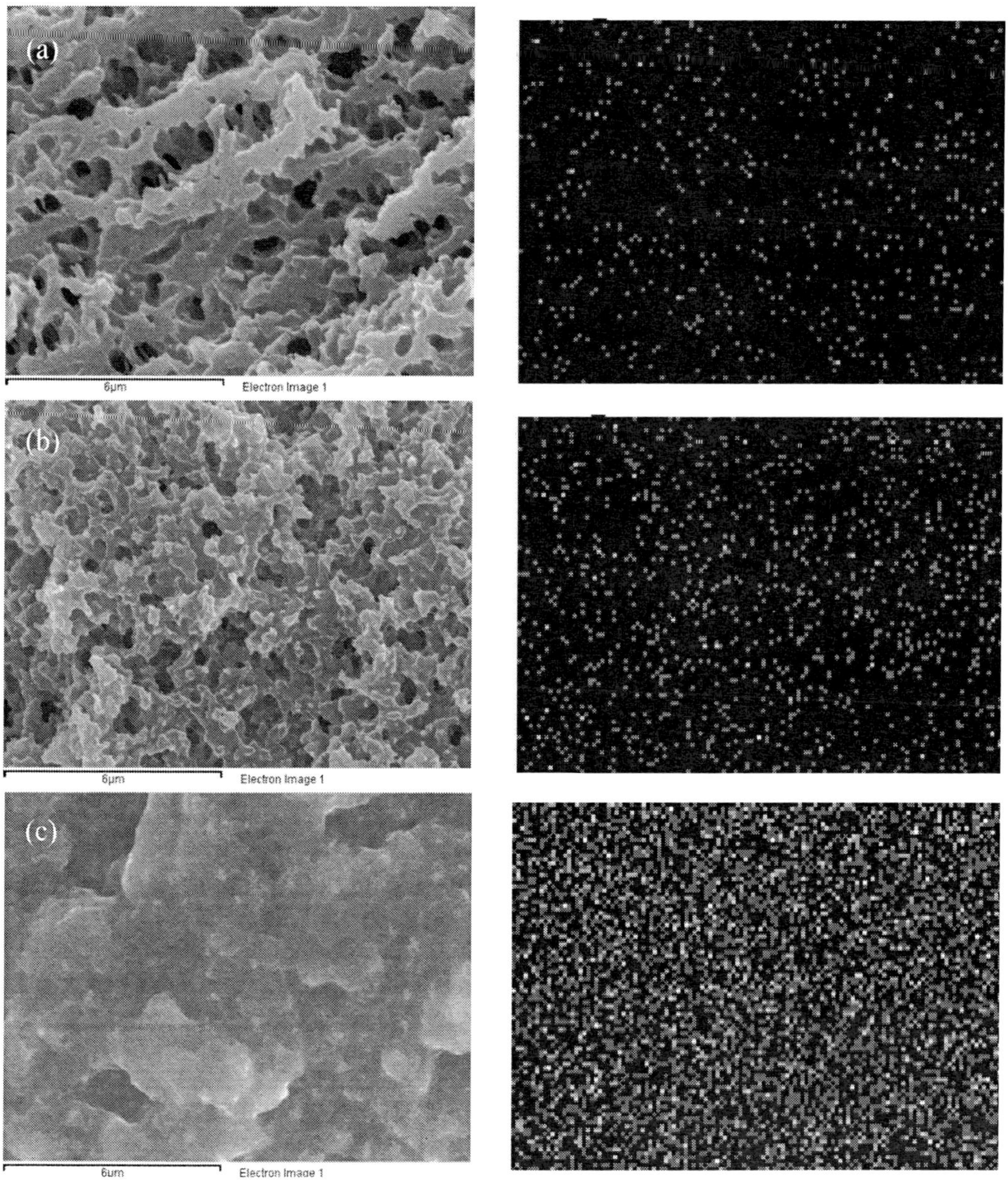

Figure 25. FE-SEM micrographs and corresponding Cl distributions of cured UF resin of the F/U mol ratio of 1.4. (a) 1% NH_4Cl, (b) 3% NH_4Cl, and (c) 10% NH_4Cl.
*Reproduced from ref. [58] with by permission from John Wiley & Sons (© John Wiley & Sons, 2011).

The produced acid remains as residual acid in the cured UF resin. The residual acid facilitates the hydrolytic degradation of UF resins, which is eventually responsible for the FE. In this section, the distribution of Cl element was assumed as residual acid in the cured UF resin although there were Cl elements from the un-reacted NH_4Cl. We expected that the Cl distribution could provide useful information on the process of its hydrolytic degradation.

Figure 24 shows FE-SEM images and Cl distributions of cured UF resins of F/U mole ratios of 1.0. All FE-SEM images of the UF resins showed spherical structures regardless of the NH_4Cl addition levels. These structures were quite similar to those hardened without adding NH_4Cl. This result indicated that the spherical structure had been formed in the UF resin of F/U mole ratio of 1.0 after its synthesis. In fact, Johns and Dunker [48] reported a similar result, and mentioned that these structures were due to colloidal particles in the UF resin of low F/U mole ratio. But, the influence of hardener to the spherical structure was not reported yet. The distribution of the Cl element was becoming denser as the NH_4Cl addition level increased. An interesting point was the fact that the Cl element distribution clearly showed an arc shape. And the frequency of arc shape increased at both 1% and 3% NH_4Cl levels, but it was infrequent at 10% NH_4Cl level (see white arrows). These results indicated that the residual acid existed on the surface of the spherical structure of cured UF resin of F/U mole ratio of 1.0. These results also suggest that the hydrolysis of cured UF resins could start at the surface of the spherical structures. In addition, these results also support the hypothesis of forming colloidal particles in UF resin proposed by Pratt *et al.* [52].

FE-SEM images and Cl distributions of the cured UF resins of F/U mole ratios of 1.4 at different NH_4Cl addition levels are presented in Figure 25. When the UF resin of F/U mole ratio of 1.4 was cured by adding 1% NH_4Cl, the cured UF resin showed irregular shape at the fracture surface (Figure 25, a). As the NH_4Cl content increased, the irregular shape of UF resin showed a more compacted morphology with much smaller empty spaces between them (Figure 25, b & c). In the cured UF resins of F/U mole ratios of 1.4, the formations of irregular empty spaces at low NH_4Cl level and compacted structure at high NH_4Cl level could be related two different phenomena. In other words, the water molecules formed by the condensation need to make empty spaces to be evaporated, while the curing of UF resin molecules need to cross-link them, leading to a compacted structure. Thus, it was believed that the morphology of cured UF resin of F/U mole ratios of 1.4 is resulted from the combination of water evaporation and curing process. So, large empty spaces formed at 1% NH_4Cl level could be due to water evaporation at a slow curing process. In other words, a slow curing process allows the water molecules to evaporated, leading to the formation of irregular morphology. However, a compacted structure at 10% NH_4Cl level could be attributed to that a faster curing process did not allow much time for the water molecules to evaporated UF resin. As expected, the abundance of Cl element was much greater than those UF resins of F/U mole of 1.0. These results could be due to greater amount of free formaldehyde in the resin, which produced much more hydrochloride acid. In general, higher F/U mole ratio resulted in a greater amount of free formaldehyde after UF resin synthesis [48]. Thus, the arc shape distribution of Cl element was hardly observed for the cured UF resin of higher F/U mole ratio.

For comparison, the amount of each chemical element for cured UF resins was also quantified. The quantitative results of each element in the cured UF resins at different F/U mole ratios as well as NH_4Cl levels are presented in Figure 26. The quantity of each element based on their weights was given because the results based on the atom percent were quite similar each other. Regardless of the F/U mole ratio, the C quantity decreased as the NH_4Cl level increased. This result could be due to the evaporation of formaldehyde in UF resins during the curing process. But, the N element showed an increased quantity with an increase in the NH_4Cl level as expected, although it slightly decreased at 1% NH_4Cl level. This could be due to the N element of the NH_4Cl added. And the quantity of O element in the cured resin decreased with an increase in the NH_4Cl level. This result might be due to attributed to the condensation reaction of hydroxymehtyl groups into methylene linkages by splitting water as a by-product. In other words, the oxygen in the mehtylol groups disappeared during the curing of the resin, which reduced the quantity of oxygen in the cured UF resin. But, the quantity of chloride increased as the NH_4Cl level increased as expected. However, the amount of Na element was within 1%, which was the least amount among the elements detected in the cured UF resins.

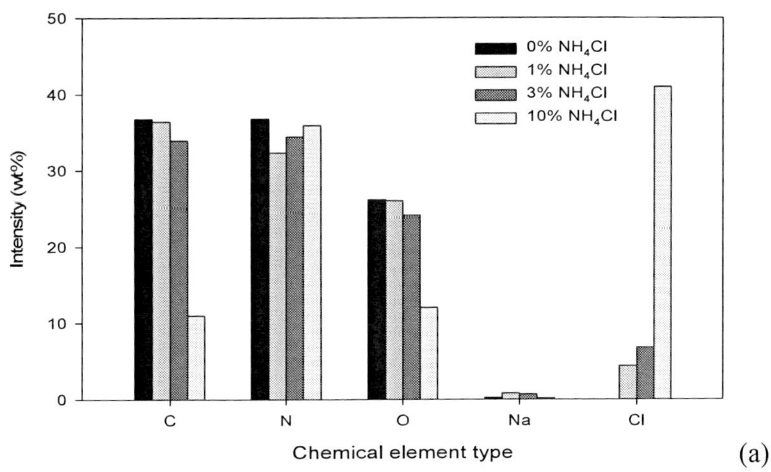

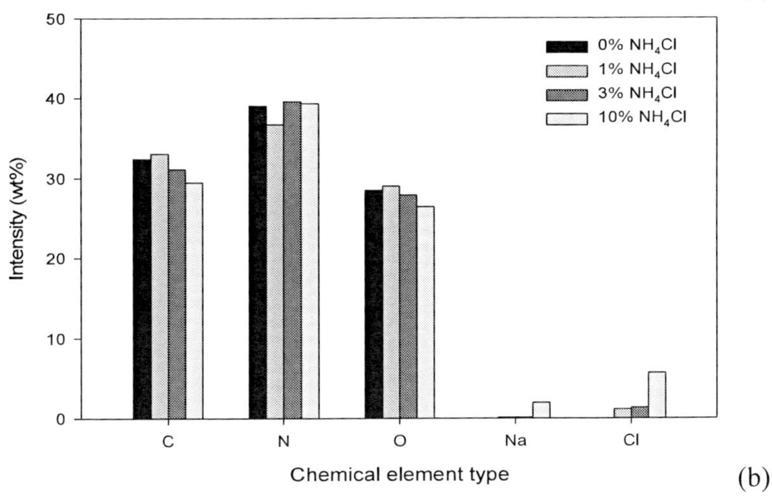

Figure 26. Quantitative results of chemical elements in cured UF resin with different F/U mole ratios. (a) F/U = 1.4, (b) F/U=1.0.

*Reproduced from ref. [58] with by permission from John Wiley & Sons (© John Wiley & Sons, 2011).

4.3. Microstructure of Cured UF Resins

To further understand the microstructure of UF resins, a thin film was prepared at two different surfaces i.e., the outer and fracture surfaces. Prior to looking at the microstructure at the outer surface of the cured UF resin films, comparison was made between the AFM image of the UF resin films prepared by adding without and with NH_4Cl as shown in Figure 27. The AFM image of the hardened UF resin films without adding NH_4Cl shows very rough surfaces with two distinctive regions, i.e., bright and dark areas (Figure 27, a). This rough surface could be the result of the spherical structures of the UF resins that have been reported for the low formaldehyde/urea mole ratio of 1.2 [47]. Lower F/U mole ratio UF resin has colloid particles, which are coalesced into clusters in the aging process. These clusters were known to form spherical structures in UF resin [48]. In the imaging by the AFM, higher forces tend to enhance the phase contrast, which is a function of the elastic and viscoelastic properties of the sample [53]. In other words, harder, less viscoelastic phases will be brighter than softer, more viscoelastic phases. Thus, the bright and dark areas were classified as the hard phase and soft phase, respectively. In addition, there were pores (black arrows) at the outer surface of the UF resin (Figure 28). These pores could be the result of the evaporation of water of formaldehyde during its curing process. When the UF resin was cured by adding 0.1% NH_4Cl based on the resin solids, the outer surface became much rougher than that of its counterpart. In fact, the R_a increased from 303.6 nm to 433.9 nm and the R_q increased from 389.5 nm to 574.2 nm when the NH_4Cl was added. The greater surface roughness of the cured UF resin films could be due to the formation of cross-links during it curing process.

In order to simulate the hydrolysis process in cured UF resin responsible for FE, the prepared UF resin films were etched by dilute hydrochloric acid in different times, ranging from 0 to 40 seconds. Typical AFM images obtained by the scans on the outer surface were shown in Figure 28. As expected, the hard and soft phases appeared for all AFM images, which were analyzed by software to obtain the hard phase area percentage.

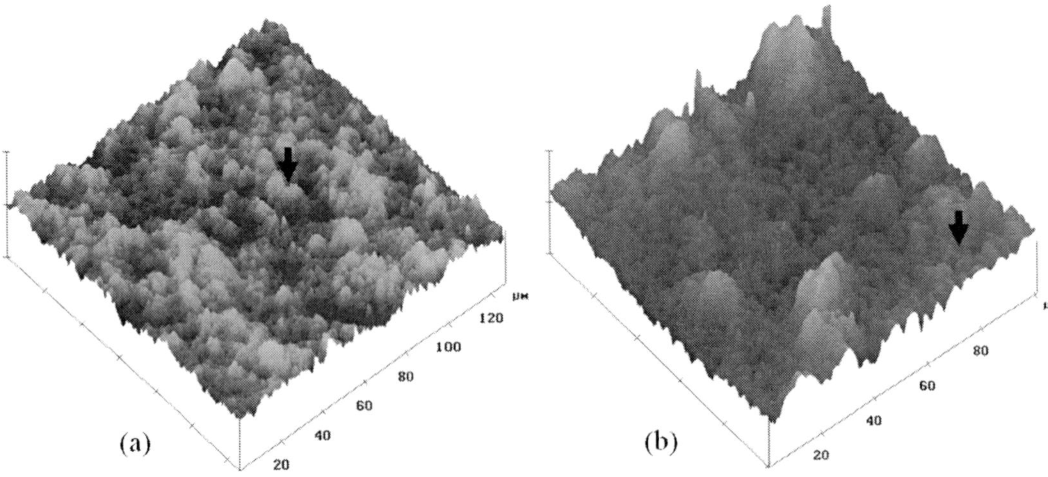

Figure 27. Outer surface AFM images of the cured UF resin film. (a) 0% NH_4Cl, and (b) 0.1% NH_4Cl. *Reproduced from ref. [54] with by permission from John Wiley & Sons (© John Wiley & Sons, 2011).

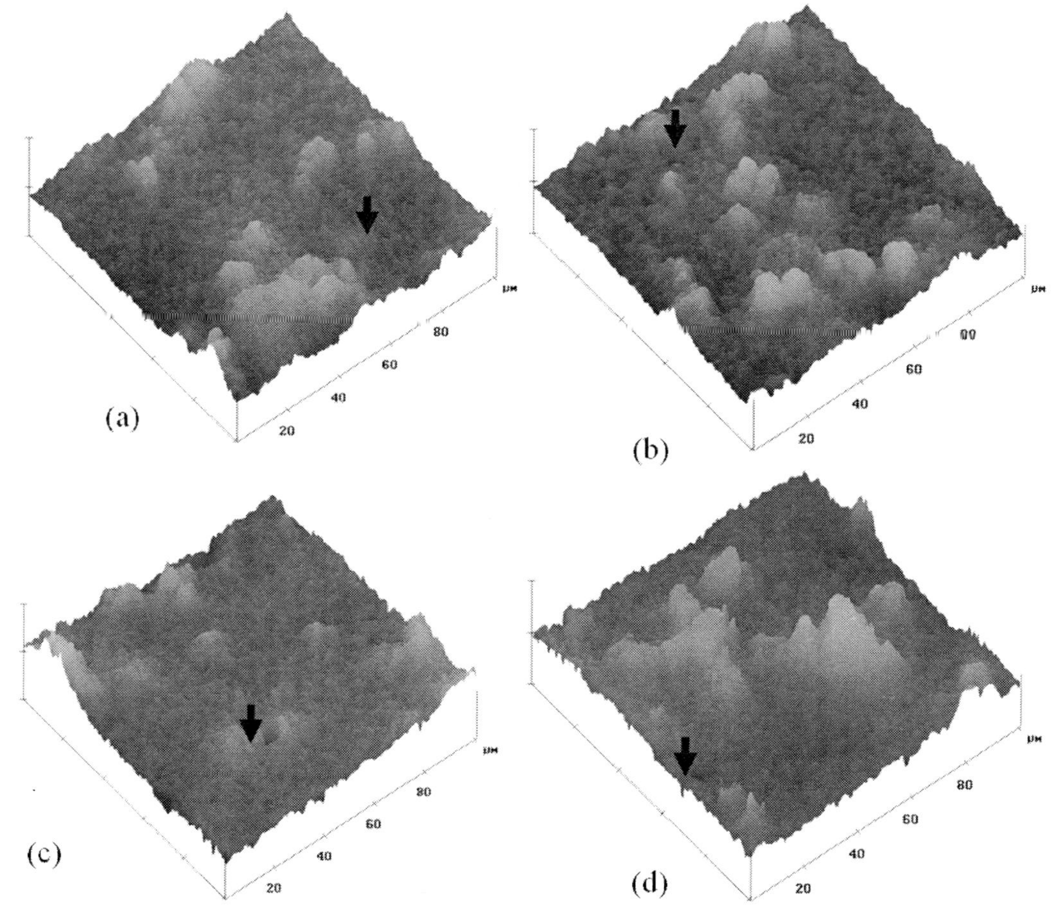

Figure 28. Outer surface AFM images of the cured UF resin film as a function of etching time. (a) 0s, (b) 10 s, (c) 20 s, and (d) 40 s.
*Reproduced from ref. [54] with by permission from John Wiley & Sons (© John Wiley & Sons, 2011).

Figure 29 (a) shows the measurement results of the hard phase areas on the outer surfaces of the cured UF resin films. Average values with standard deviation for each etching time were presented. As the etching time increased, the hard phase area increased up to 20 seconds, and then decreased with an etching time of 40 seconds. These results indicate that the hard phase area increases with an increase in the etching time because the soft phase area decreases due to hydrolytic degradation of the cured UF resin by the acid etching. However, the decreased hard phase area with an etching time of 40 seconds could also be due to the hydrolysis of both the hard and soft phases at the same time. In other words, the hydrolysis of cured UF resin starts at the soft phase first and then moves on to the hard phase. But it is believed that longer etching times result in the simultaneous removal of both the soft and hard phases. It is interesting to note that the standard deviation of the measurements increase with an increase in the etching time. This suggests that the outer surface of the cured UF resin films become rougher as each etching time is extended because the hydrolysis exposes a new hard phase area during the etching process.

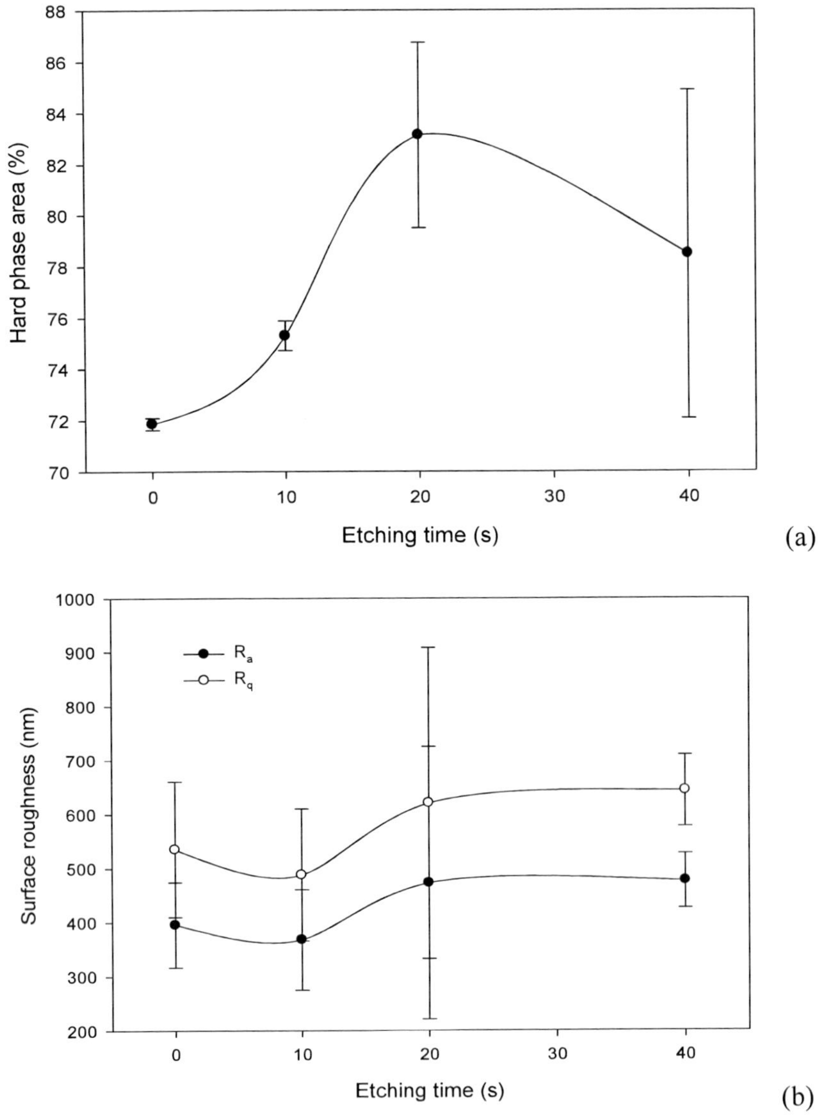

Figure 29. Surface properties of the cured UF resin films as a function of etching time. (a) Hard phase area and (b) outer surface roughness.
*Reproduced from ref. [54] with by permission from John Wiley & Sons (© John Wiley & Sons, 2011).

So, the surface roughness values of the samples are presented in Figure 29 (b). As the etching time increased, the surface roughness expressed by R_a and R_q slightly decreased at an etching time of 10 seconds, and then increased at 20 seconds of etching followed by a level-off. After 10 seconds of etching, the surface roughness had not much changed even though the hard phase area increased. But, the surface roughness increased with 20 seconds, which was consistent with the measurements of the hard phase area as shown in Figure 29 (b). These results also suggest that the hydrolysis simulated by acid etching removes the soft phase first and then hard phase.

To compare the outer surface's microstructure, the AFM was also applied to the fracture surface of cured UF resin by adding 0.1% NH_4Cl. The AFM image is shown in Figure 30, (a). As expected, the AFM image also showed the hard and soft phases. The AFM scans also found a pore (black arrow in Figure 30, a), which was believed to be formed by the evaporation of water during its curing process. An interesting point is that the hard phase, i.e., brighter region, shows filament-like sharp structures (white arrows) with various dimensions (Figure 30, a). These brighter structures in the AFM images mean a lot of hard and less viscoelastic regions. Thus, these are believed to be crystalline structures in the cured UF resins.

For the first time, the author reported the three dimensional shapes of the filament-like thin structures on the fracture surface of the cured UF resins [54]. Although the reason is not clear, the formation of these structures could be due to the presence of colloids of the UF resins with low F/U mole ratios. In other words, it was reported that filament-like colloidal aggregates were initially formed in UF resin, and then eventually changed to super-clusters by the coalescence in later during the ageing process [47-48]. And the presence of crystalline structures in UF resin has been reported by several authors [52, 55-57].

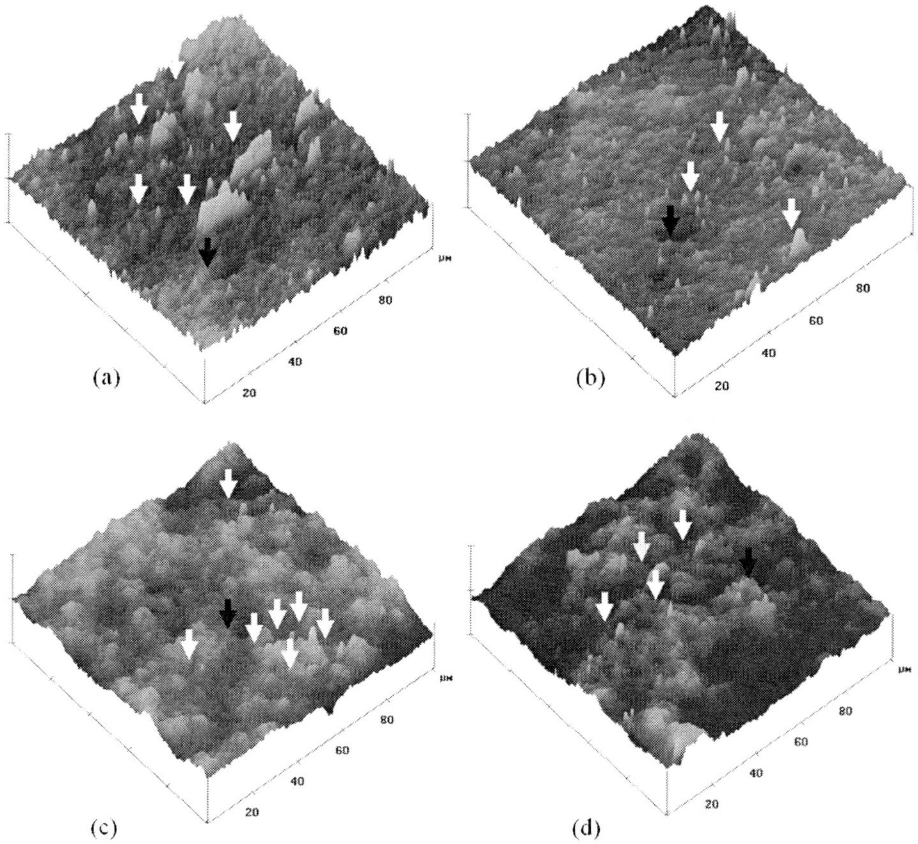

Figure 30. Fracture surface AFM images of the cured UF resin before and after etching. (a) and (c) hardened by ammonium chloride before etching, (b) and (d) hardened by ammonium sulfate after etching for 10 s.
*Reproduced from ref. [54] with by permission from John Wiley & Sons (© John Wiley & Sons, 2011).

In particular, UF resins with lower F/U mole ratio of 1.2 are supposed to have colloidal structures, which constitute crystalline structure [48, 56]. Dunker *et al.* [48] reported that UF resin contained colloidal regions of semicrystalline nature, and ascribed the origin of the crystal structure to a high degree of order due to hydrogen bonding. They also mentioned the possibility that the crystalline regions could have arisen from the crystallization of some minority components of UF resin, like urons or other ring structures. A physical association in the UF resin solution was related to the crystalline region in solid form [52].

However, it is not certain what process is playing a role in the formation of these filament-like thin structures. Further research is required to find more conclusive evidence on how the filament-like thin structures are formed by the colloidal particles in the UF resins.

The facture surface was etched by dilute hydrochloric acid for 10 seconds, and then the AFM image obtained was presented in Figure 30, (b). As expected, both the hard and soft phases were observed, and the filament-like thin structures of the hard phase was also detected. In general, the fracture surface became much smoother after the acid etching for 10 seconds than before the etching. Also, the size of the hard phase's thin structure decreased after the acid etching. Although the size of the filament-like thin structure decreased, the occurrence frequency of the structure increased greatly. This phenomenon could be ascribed to the hydrolysis process in that the acid removed either the hard phase, showing the remaining crystalline structures with a decreased size; or the soft phase, emerging new the crystalline structures at the fracture surface. Regardless of the degradation behavior, these results clearly indicate that the hydrolysis of UF resin caused by etching degrades the hard and soft phases.

To compare the microstructure of cured UF resins depending on the type of hardener, AFM images of cured UF resins by adding ammonium sulfate were also presented in Figure 30. As expected, the pore for water evaporation (black arrow) was also observed for both the control and etched samples. When the UF resin was cured by adding ammonium sulfate, the topography of the facture surface was quite different from that of the cured UF resin by adding ammonium chloride (Figure 30, c). In other words, a limited number of the crystalline structures (white arrows) occurred even though the coverage of the brighter hard phase was apparently much larger. The acid etching of the cured UF resin by ammonium sulfate exposed a greater number of the crystalline structures on the facture surface (Figure 30, d). This result indicates that the molecular structure of UF resins cured by ammonium sulfate is much more resistant to hydrolysis than those cured by ammonium chloride. This result is quite compatible with the measurement of the hydrolytic stability of cured UF resins [47]. The authors reported that cured UF resins by ammonium sulfate chloride had greater hydrolytic stability than cured UF resins by ammonium chloride.

The AFM scans of the fracture surface of the cured UF resins also provided two different surface roughness results (i.e., R_a and R_q) as shown in Figure 31. The surface roughness values of the cured UF resins by ammonium chloride after the etching decreased as presented in Figure 31, a. As discussed in the previous section, this result could be due to the hydrolysis of the cured UF resin by ammonium chloride. In other words, the surface roughness decreases because the hydrolysis due to etching simultaneously degrades the hard phase area with crystalline structure as well as the soft phase area as shown in Figure 31. However, the surface roughness increased after the etching for the cured UF resin by ammonium sulfate as presented in Figure 31. In fact, the R_a value increased from 279.2 nm to 312.3 nm and the R_q value increased from 350.8 nm to 393.2 nm. These results suggest that the hydrolysis of UF resins cured by ammonium sulfate dominantly degrades the soft phase area and exposes a greater number of crystalline structures on the fracture surface as discussed in Figure 31. A greater susceptibility to the hydrolytic degradation of UF resins cured by ammonium chloride than those by ammonium sulfate also supports the observed hydrolysis behavior by the AFM, depending on the hardener types [56].

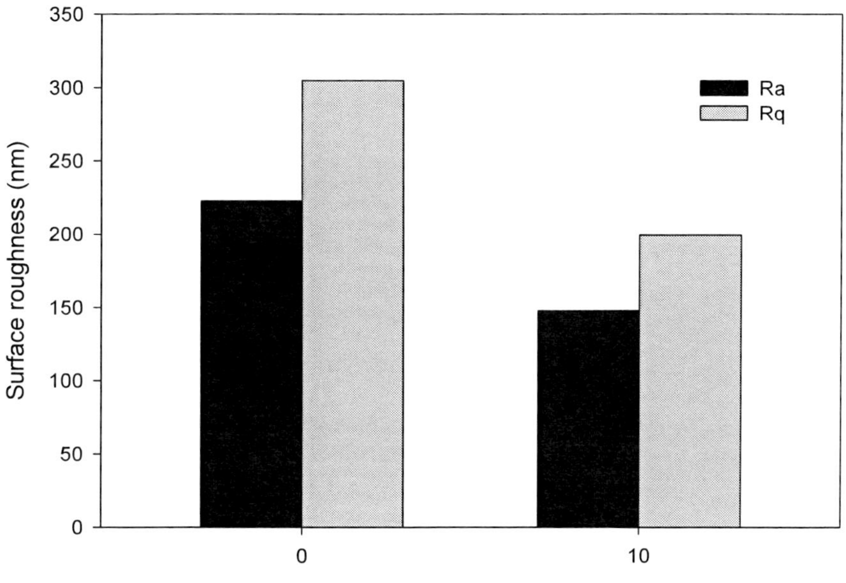

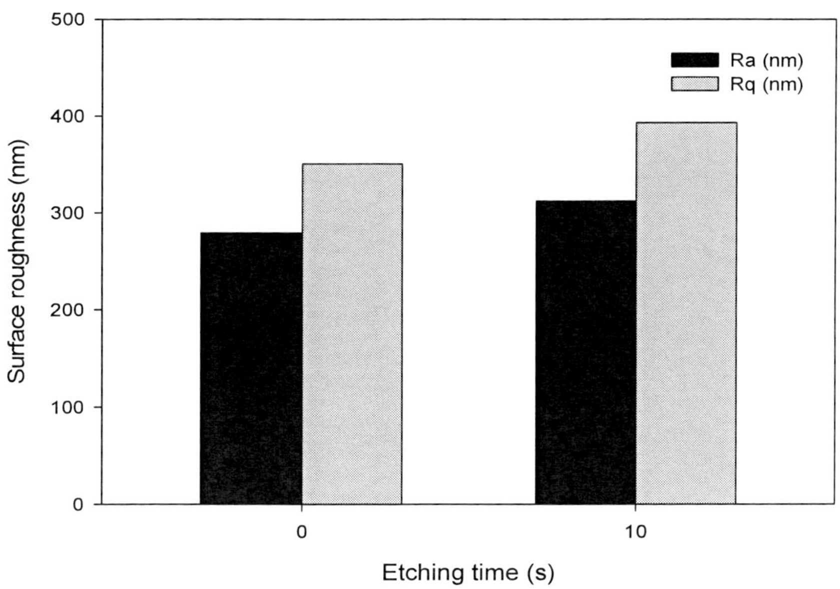

Figure 31. Fracture surface roughness of the cured UF resin film by adding (a) ammonium chloride and (b) ammonium sulfate as a function of etching time *Reproduced from ref. [54] with by permission from John Wiley & Sons (© John Wiley & Sons, 2011).

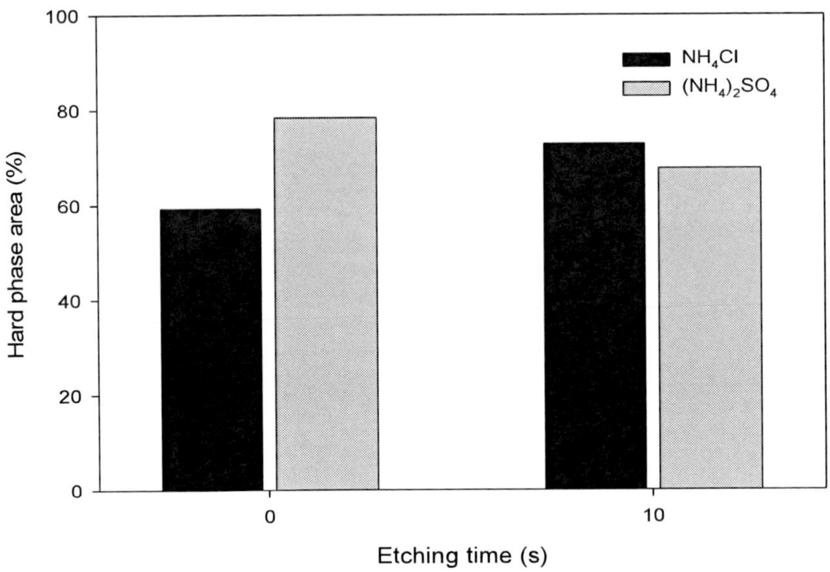

Figure 32. Hard phase area of the cured UF resins by adding ammonium chloride and ammonium sulfate as a function of etching time.
*Reproduced from ref. [54] with by permission from John Wiley & Sons (© John Wiley & Sons, 2011).

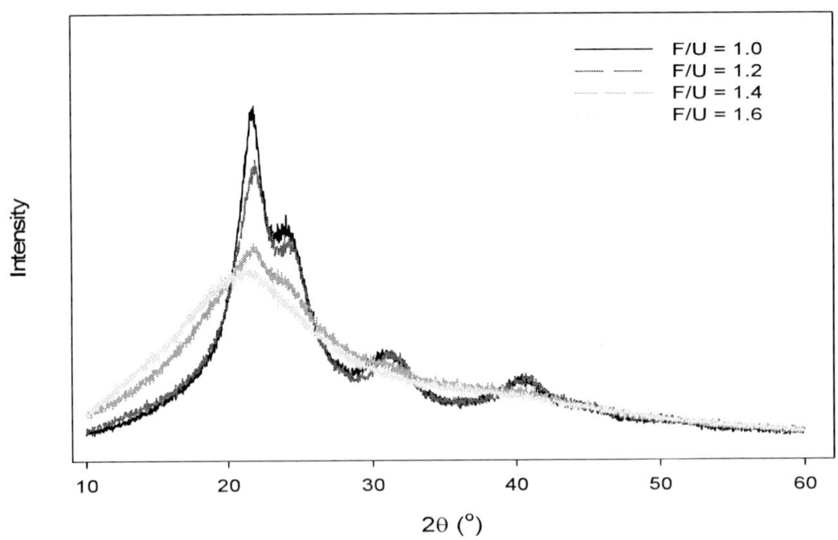

Figure 33. X-ray diffractograms of cured UF resins with different F/U mole ratios. Curing condition: 3% NH_4Cl, 120°C, 60 min.
*Reproduced from ref. [59] with by permission from Springer (© Springer, 2011).

In addition, the hard phase area percentage obtained by image analysis of the AFM images before and after the etching is presented in Figure 32. The percentage of the hard phase area increased from 59.3 % to 72.8 % after the etching of the fracture surface of UF resins cured by adding ammonium chloride. As mentioned in the previous paragraphs, an increase in the hard phase area after the etching could be the result of a greater susceptibility to the hydrolysis of ammonium chloride cured UF resins. In other words, the soft phase of cured UF resins by ammonium chloride was easily hydrolyzed, which resulted in a greater area of the hard phase being exposed after the etching. But, the hard phase area slightly decreased after the etching when the UF resin was cured by adding ammonium sulfate. This could be due to a greater resistance to hydrolysis of the cured UF resins by ammonium sulfate. The soft phase was slightly degraded by the hydrolysis caused by the etching, which reduced the hard phase area after the etching.

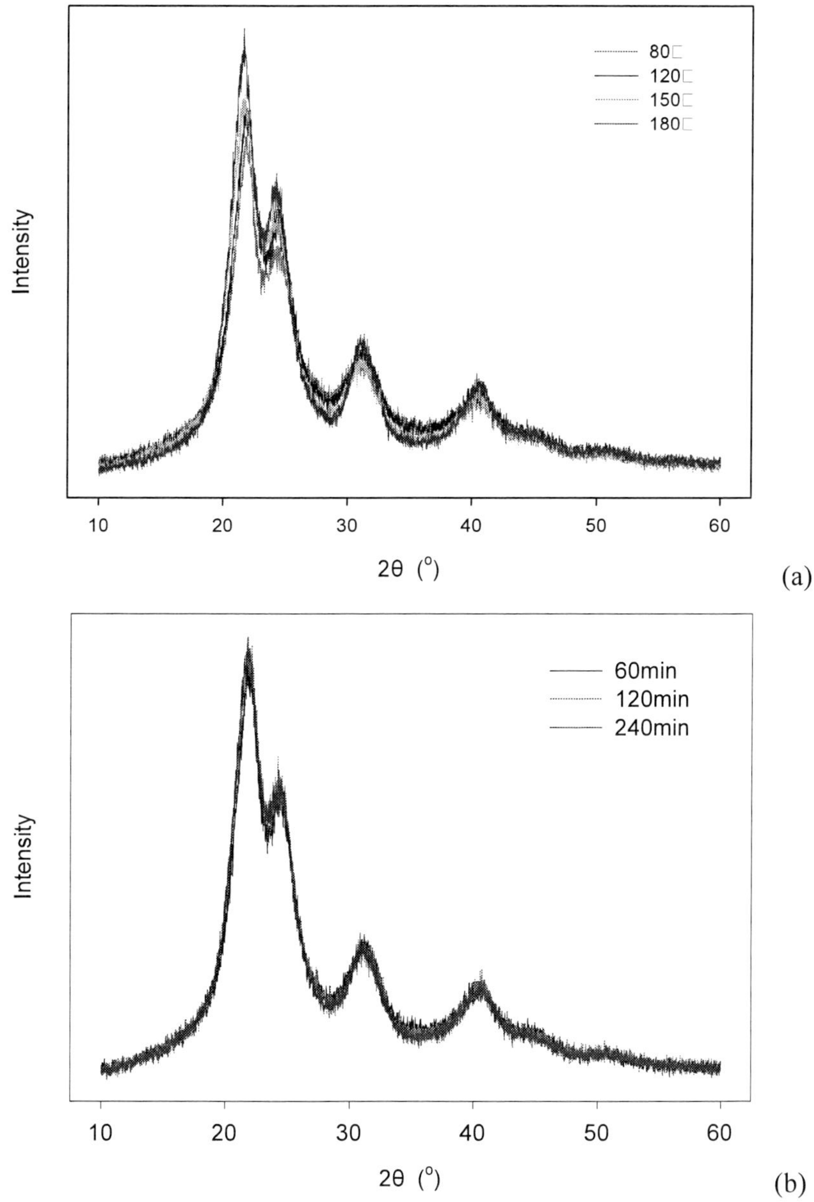

Figure 34. X-ray diffractograms of the cured UF resins with the F/U mole ratio of 1.0 at different curing (a) temperatures, and (b) times. Curing condition: 3% NH_4Cl, 120°C, 60 min.
*Reproduced from ref. [59] with by permission from Springer (© Springer, 2011).

4.4. Crystalline Structure of Cured UF Resins

Figure 33 displays the results of x-ray diffractograms of cured UF resins, depending on F/U mole ratios. As shown, the UF resins with the F/U mole ratios of 1.6 had a single main peak at a 2θ of about 21°. By contrast, that of cured UF resin with the F/U mole ratio of 1.4 showed the same strong peak as well as a weak peak around its shoulder. These results indicate that the cured UF resins with higher F/U mole ratios are amorphous structure [25]. When the F/U mole ratio further decreased to 1.2 and 1.0, these two peaks became sharper and showed much greater intensity. In addition, two additional peaks that appeared at about 31° and 40° indicate additional crystalline regions for these cured UF resins. Each of the d-spacing was calculated as 2.86 Å and 2.22 Å. These results are consistent with other reported results [47-48, 52, 55-56]. In other words, UF resins at lower F/U mole ratio from 1.1 to 0.5 shows crystal structure while UF resins with higher F/U mole ratio are amorphous polymer. For example, Levendis *et al.* [56] mentioned that UF resins at lower F/U mole ratio than 1.1 showed crystal structure while UF resin with higher F/U mole ratio than 1.5 were amorphous structure. However, the result shows that UF resin with a lower F/U mole ratio of 1.2 also possess crystalline regions, which are observed for the first time.

In addition, these results are quite interesting in terms of the hydrolytic stability of cured UF resins at these low F/U mole ratios. In general, cured UF resins at lower F/U mole ratios of 1.2 and 1.0 had a greater hydrolytic stability and these resins showed additional crystalline regions at the same time. These results suggest that an improved hydrolytic stability of cured UF resins with lower F/U mole ratios of 1.2 and 1.0 be related to the additional crystalline regions of the resins. In other words, the crystalline regions could provide more resistance to the degradation of their hydrolysis process than those of higher F/U mole ratio resins.

In order to understand whether these crystalline regions of cured UF resin with the F/U mole ratio of 1.0 are inherent, these crystalline regions of cured UF resin with the F/U mole ratio of 1.0, depending on curing conditions, hardener type are studied. Figure 34 shows the X-ray diffractograms of the UF resin of F/U mole ratio of 1.0 which was cured at different curing temperatures and times. The intensities of all crystalline regions increased as the curing temperature increased (Figure 34, a). This result indicates that higher curing temperature causes a greater amount of crystalline regions of the UF resin. But the 2θ position of two additional peaks did not change with the curing temperature. This result indicates that cured UF resins with lower F/U mole ratios (1.2 and 1.0) had additional crystalline structures. It was reported that these low F/U mole ratio UF resins was reported to have spherical structures [58]. The authors believed that these additional crystalline peaks could be related to the spherical structures observed at the UF resins of lower F/U mole ratios. X-ray diffractograms of the UF resins depending on curing time are shown in Figure 34 (b). As the curing time increases, the peak intensities of the two additional peaks also increase. As the curing temperatures, this result also suggests that higher curing temperature causes a greater amount of the crystalline regions of the resin.

Influences of different hardener types and addition levels are also investigated for the UF resin, which are presented in Figure 35. No hardener samples (or 0% NH_4Cl) were prepared by hardening the sample by drying at 120°C overnight to remove the water content of the resin. As shown in Figure 35 (a), the intensities of the crystalline regions of the UF resin increased up to 3% hardener level and then slightly decreased for three hardener types. These results suggest that the addition level of hardener influences the extent of the crystalline regions, not on the peak position. This result is partially consistent with the reported results that the crystallinity increased with an increase in the NH_4Cl level and curing temperature [56]. However, the types of hardener did not affect the additional crystalline regions of the UF resin. Figure 35 (b) and (c) shows X-ray diffractograms of cured UF resins by adding ammonium sulphate, aluminium sulphate at different levels. Two hardeners also showed a quite similar trend to that of the ammonium chloride. These results indicated that hardener type and its addition level did not change the additional crystalline regions of the UF resin. Even though the curing condition, hardener type and level influenced the intensity of the crystalline regions, the crystalline peak positions did not change, depending on these parameters. This result indicates that the crystalline regions are inherently present for the cured UF resin with the F/U mole ratio of 1.0 [59].

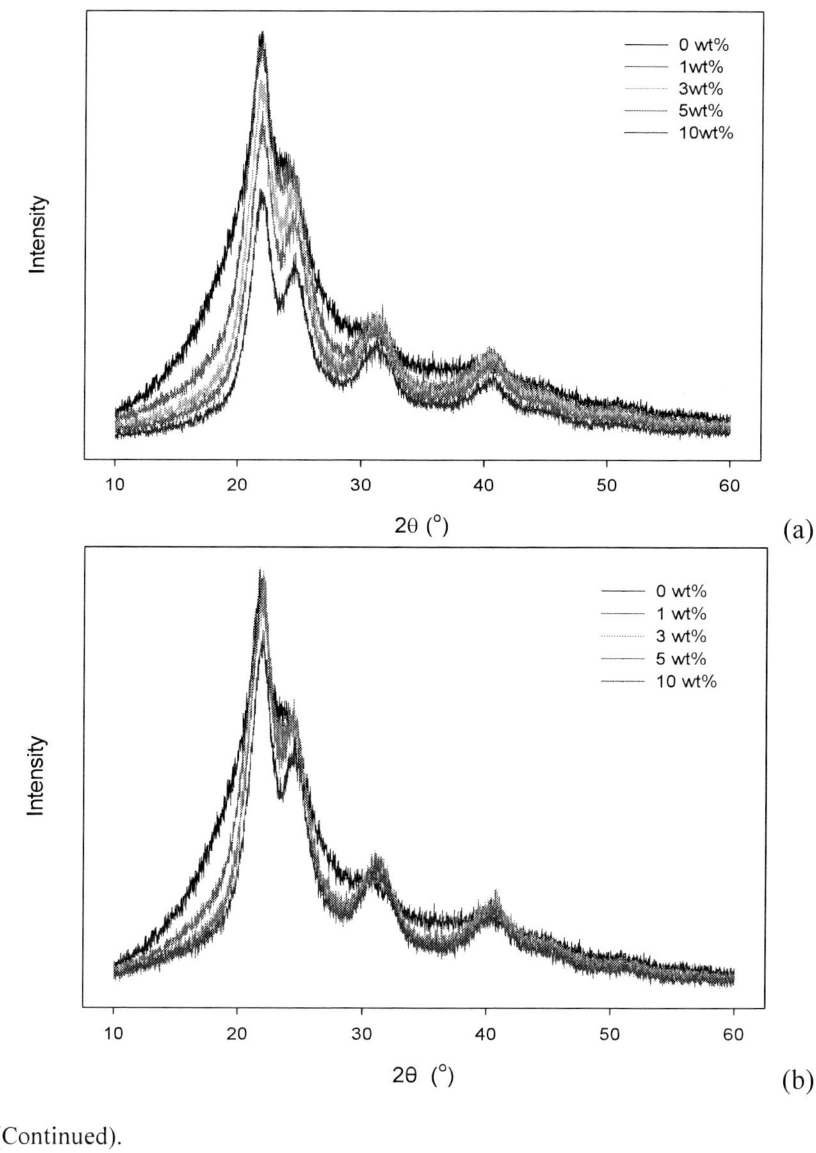

Figure 35. (Continued).

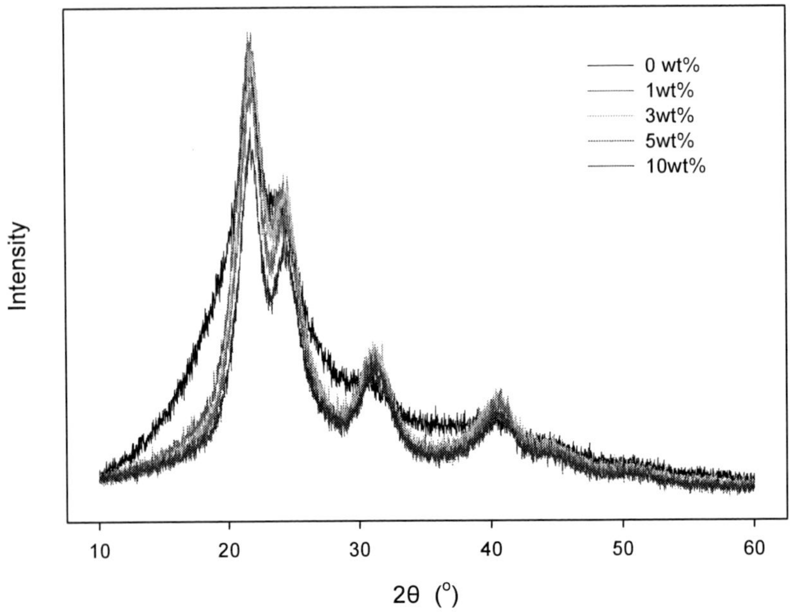

(c)

Figure 35. X-ray diffractograms of cured UF resins of F/U mole ratio of 1.0 as a function of hardener type and level. (a) NH_4Cl, (b) $(NH_4)_2SO_4$, and (c) $(Al_2)(SO_4)_3$. Curing condition: 120°C, 60 min. *Reproduced from ref. [59] with by permission from Springer (© Springer, 2011).

Figure 36 exhibits X-ray diffractograms of solid urea, monomethylol urea and dimethylol urea as well as that of the cured UF resin with the F/U mole ratio of 1.0. Solid urea clearly illustrates crystalline structures in the X-ray diffractogram (Figure 36, a). The X-ray diffractogram of the monomethylol urea is quite different from that of the solid urea (Figure 36, b). And the peaks of additional crystalline regions did not overlap with those of the monomethylol urea. These results indicate that the crystal structure of monomethylol urea does not contribute much as that of the cured UF resin. The X-ray diffractograms of both dimethylol urea and that of the cured UF resin are shown in Figure 36, (c). The X-ray diffractogram of the dimethylol urea is quite similar to that of the cured UF resin, which suggests that the crystalline regions of the cured UF resin with the F/U mole ratio of 1.0 are mainly influenced by the presence of dimethylol urea. In other words, the crystalline regions of the cured UF resins with the F/U mole ratios of 1.2 and 1.0 could be mainly composed of dimethylol ureas.

5. Modification of UF Resins

5.1. Formaldehyde Scavengers Modified UF Resins

In order to abate the FE of UF resin-bonded wood-based composite panels, two formaldehyde scavengers, i.e., urea-formaldehyde prepolymer (UFP) and urea solution (US) are blended with neat UF resins. Two scavengers prepared were mixed with UF resin at different proportions (100:0, 90:10, 80:20, 70:30, 50:50), on the basis of non-volatile solid content, which resulted in different non-volatile solids contents of the modified UF resins, as shown in Table 10.

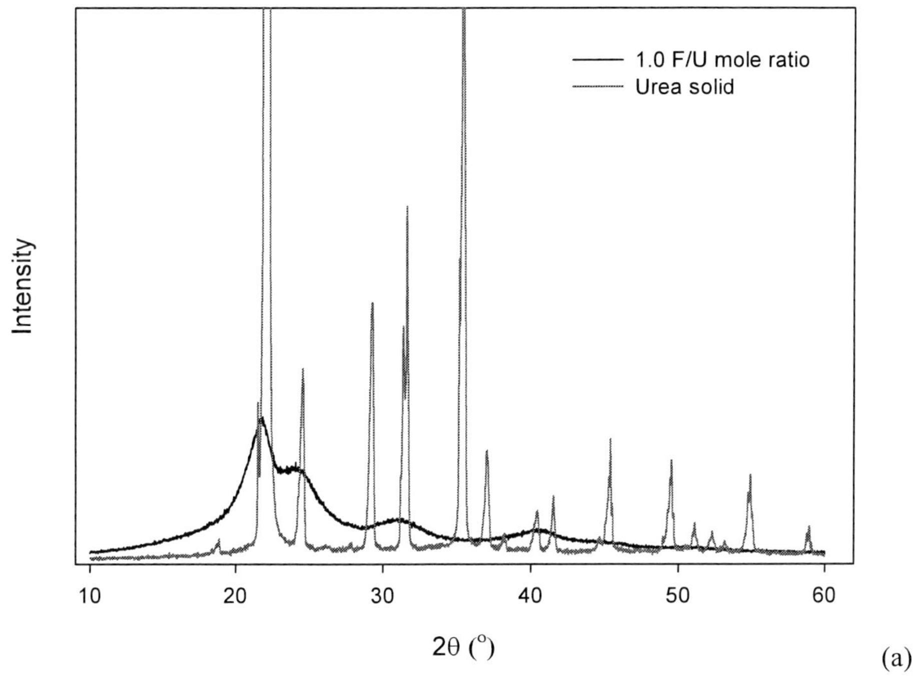

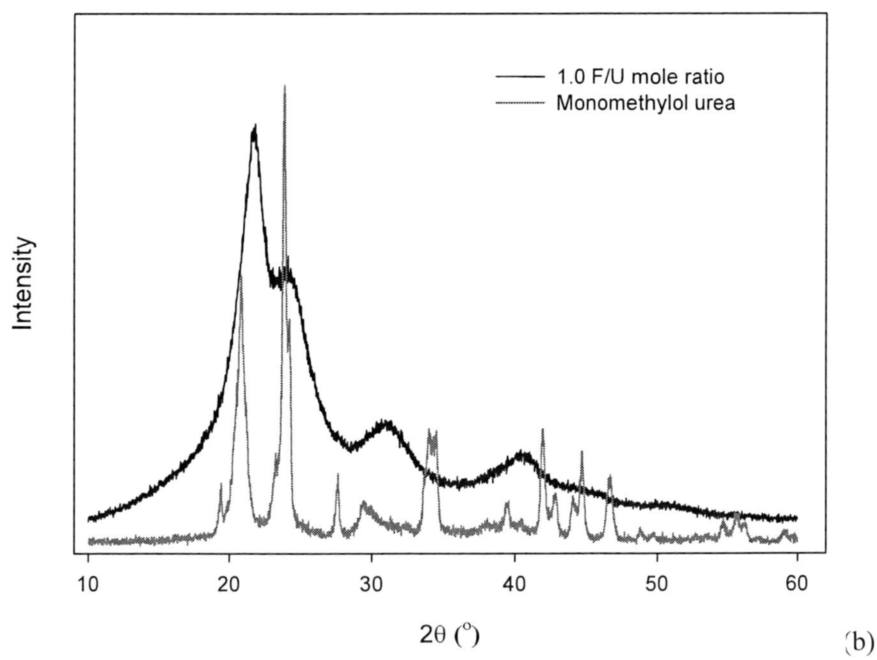

Figure 36. (Continued).

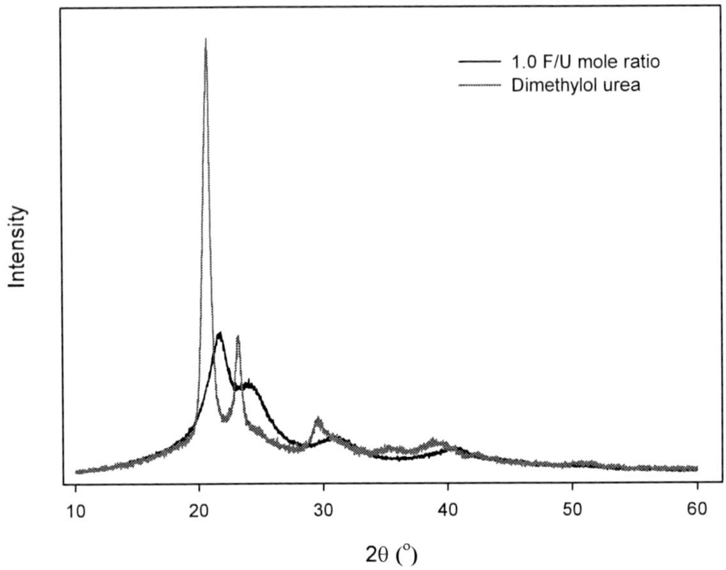

(c)

Figure 36. Comparisons of X-ray diffractograms cured UF resins (1.0 F/U mole ratio) with (a) solid urea, (b) monomethylol urea, and (c) dimethylol urea. Curing condition: 3% NH_4Cl, 120°C, 60 min. *Reproduced from ref. [59] with by permission from Springer (© Springer, 2011).

Figure 37 shows the gel time of modified UF resins with UFP or US. As the scavenger portion in the UF resin increases, the gel time did not change much for the UFP modification while it gradually increased to 30wt% of the US addition and then it dramatically increased to 50 wt% of the US addition, but the gel time of the UF resin modified with 50 wt% UFP slightly increased (Figure 37). Within the range of the scavenger concentrations, the gel time of the UF resin was greater for UFP than that of US. These results suggest that the reactivity of the modified UF resins with UFP is greater than those resins with US. This result could be attributed to the presence of methylolated ureas that were formed in the UFP, which are a product of the initial addition reaction between formaldehyde and urea under alkaline condition. In other words, methylolated ureas are able to subsequently form methylene linkages during the curing of modified UF resin. By contrast, the urea in the US does not possess methylolated ureas, which requires more time to cure the modified UF resin as the concentration of US increases.

Table 10. Non-volatile solids content of modified UF resins, depending on the various concentrations of the two scavengers (i.e., UFP and US)

Scavenger Type	Mixing Ratio (UF : Scavenger, wt %)				
	100:0	90 : 10	80 : 20	70 : 30	50 : 50
UFP	60.4	58.4	56.5	54.8	50.7
US	60.4	58.2	54.8	51.9	48.5

Reproduced from ref. [66] with by permission from John Wiley & Sons (© John Wiley & Sons, 2008).

Table 11. Properties of particleboards bonded with modified-UF resins by the addition of various concentrations of the two formaldehyde scavengers

Properties	Scavenger type	Scavenger mixing ratio (UF resin : scavenger)				
		100:0	90 : 10	80 : 20	70 : 30	50 : 50
MOR (kgf/cm^2)	UFP	176.4	178.3	155.1	119.4	71.9
	US		179.7	134.0	75.5	-
MOE ($\times 10^3$ kgf/cm^2)	UFP	29.9	34.4	33.8	32.0	25.1
	US		35.1	30.0	16.5	
Thickness swelling (%)	UFP	8.18	7.4	9.2	14.5	42.8
	US		7.2	10.9	27.1	-
Water absorption (%)	UFP	31.3	28.9	32.5	40.1	77.2
	US		23.8	32.1	50.7	-

* Reproduced from ref. [66] with by permission from John Wiley & Sons (© John Wiley & Sons, 2008).

Table 12. Properties of modified UF resins at different levels of acrylamide

Acrylamide level (%wt)	Non-volatile solid content (%wt)	Viscosity (25°C, mPa.s)	Free HCHO (%)	Gel time (s)
0	55.20	225.0	1.27	234
1	62.05	268.0	1.11	120
4	62.10	263.0	1.15	135
6	63.91	278.0	1.16	140

Reproduced from ref. [66] with by permission from John Wiley & Sons (© John Wiley & Sons, 2008).

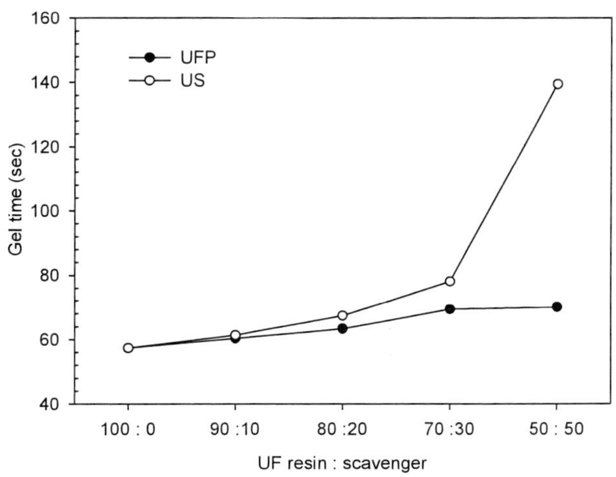

Figure 37. The gel times of modified-UF resins with two different scavengers.
*Reproduced from ref. [66] with by permission from John Wiley & Sons (© John Wiley & Sons, 2008).

Thermal curing parameters such as onset temperature, peak temperature, heart reaction, and rate constants of the modified-UF resins with either UFP or US at different scavenger concentrations are shown Figure 38. First, the onset temperatures of the modified UF resins are illustrated in Figure 38 (a). The onset temperature is defined as temperature that is linearly extrapolated by any transition or phase change as determined from a DSC curve. Thus, the onset temperature may be expressed as an extrapolated and starting temperature of the curing of the UF resin under acidic condition. As the concentration of the two scavengers increased, the onset temperature continuously increased from about 83.2°C to 89°C for the US-modified UF resins, while it slightly decreased from about 83.2°C to 82.3°C, except the addition of 30 wt% UFP. The previous study showed that the onset temperature of the UF resin depended on the F/U mole ratio, ranging from about 84.1°C to 72.6°C for the F/U mole ratio from 1.0 to 1.6 [40]. The onset temperature of the neat UF resin in this work was within the range, while the onset temperatures of the modified UF resins were much greater than those of the previous study. This result indicates that UFP-modified UF resin has a faster start of polymerization than the US-modified UF resin. In other words, the addition of US into UF resin retarded the initiation of curing, while the addition of UFP slightly accelerated the initiation of curing. Again, this result is possibly due to the presence of different chemical species in the scavenger. UFP possesses reactive methylolated ureas while the US does not.

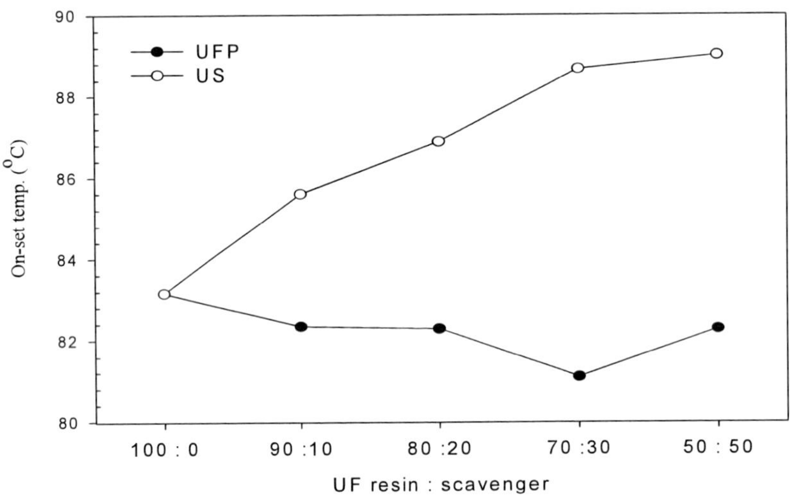

(a)

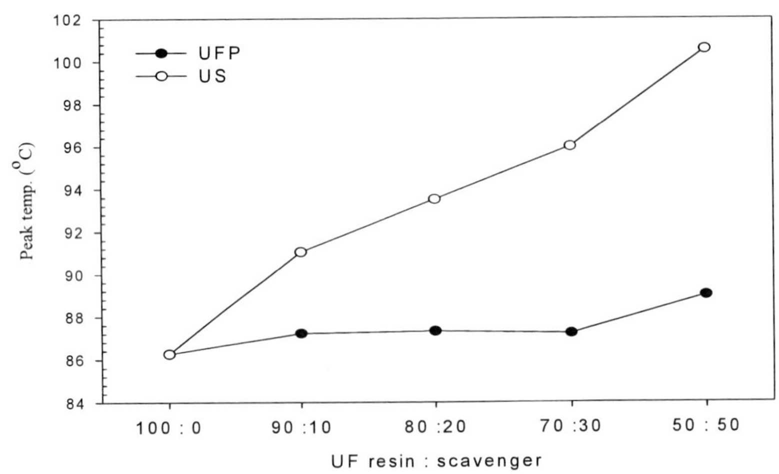

(b)

Figure 38. (Continued).

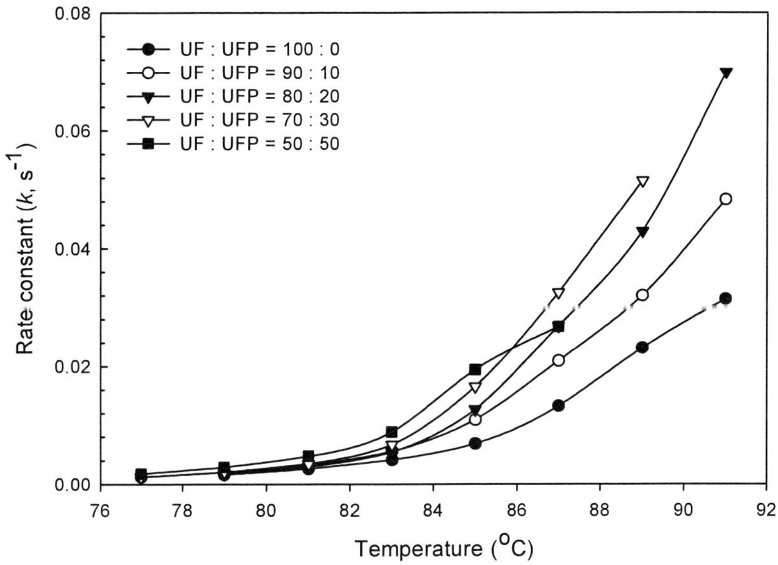

(c)

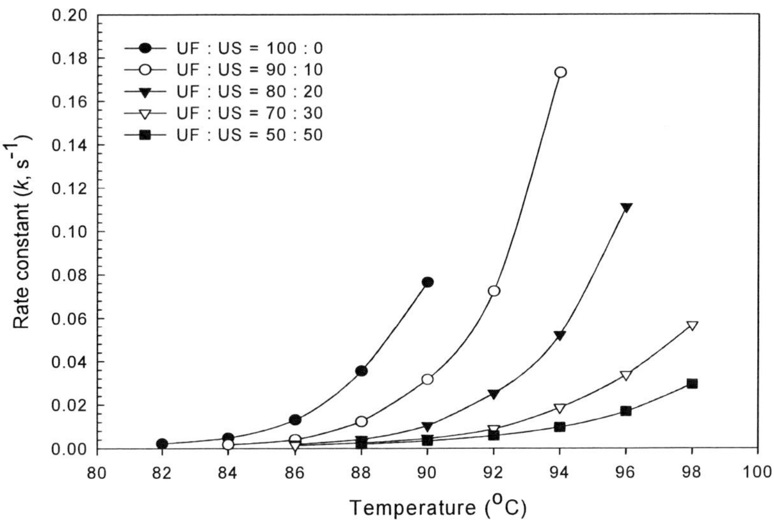

(d)

Figure 38. Thermal curing parameters of modified-UF resins with two different scavengers. (a) On-set temperature, (b) peak temperature, (c) rate constants of UFP modified-UF resins, and (d) rate constants of US modified-UF resins.

*Reproduced from ref. [66] with by permission from John Wiley & Sons (© John Wiley & Sons, 2008).

The peak temperature, where the conversion rate of UF resin reached a maximum, is presented in Figure 38 (b) which shows changes in the peak temperature of the modified UF resins. This parameter also indicates the reactivity of UF resin as does the gel time. As the scavenger concentration increases, the peak temperature of the UFP-modified UF resin did not change much, except with the addition of 50 wt% UFP, while that of the US-modified UF resin continuously increased. The US-modified UF resin showed a greater increase in its peak temperature than that of the UFP-modified one. These results show that the addition of US into the UF resin reduced the reactivity of UF resin much more than that of UFP. In other words, the addition of UFP maintained the reactivity of the UF resin, except with the addition of 50 wt%. This result indicates that the addition of UFP is more effective than that of US in simultaneously scavenging formaldehyde and maintaining UF resin reactivity.

In order to compare the reactivity of the modified UF resins with two different scavengers, the rate constants of the modified UF resins, depending on the scavenger concentration, are presented in Figure 38 (c) and (d). In general, the UF resin rate constant increased as the UFP concentration and temperature increased. As expected, this result is compatible with those of gel time, onset temperature, and peak temperature. Figure 38 (d) shows the rate constant of the US-modified UF resin as a function of temperature and US concentration. As shown, the rate constant increased with an increase in temperature, but it decreased with an increase in the concentration of US in the UF resin. This result explains the occurrence of higher peak temperatures of UF resins when they were modified with the addition of US. In other words, a lower rate constant resulted in a higher peak temperature for US-modified UF resins.

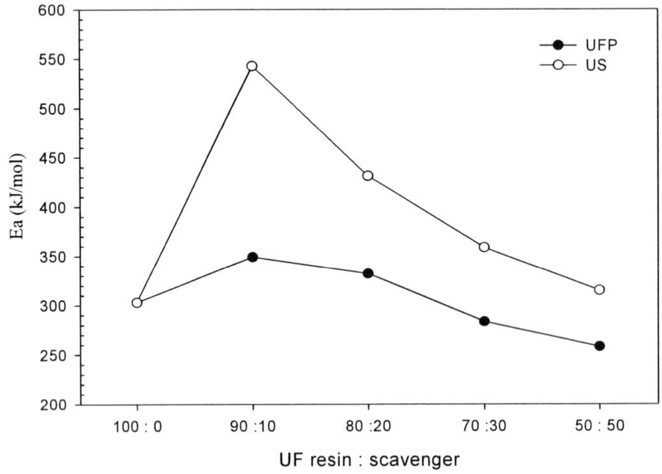

(a)

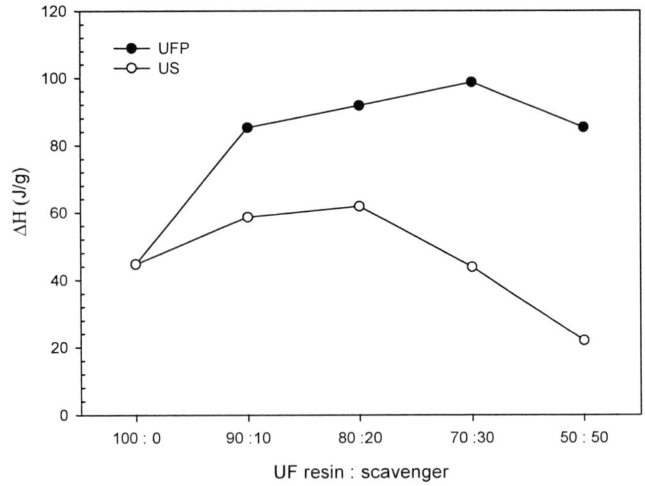

(b)

Figure 39. Thermal properties of modified-UF resins with two different scavengers. (a) Activation energy, (b) heat of reaction.
*Reproduced from ref. [66] with by permission from John Wiley & Sons (© John Wiley & Sons, 2008).

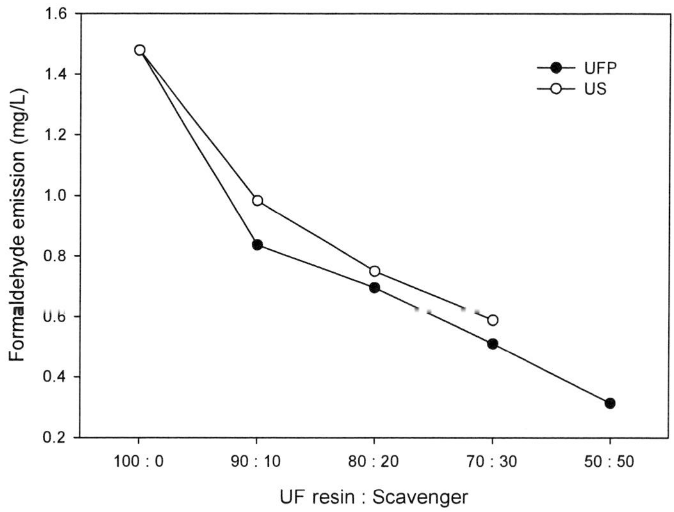

(a)

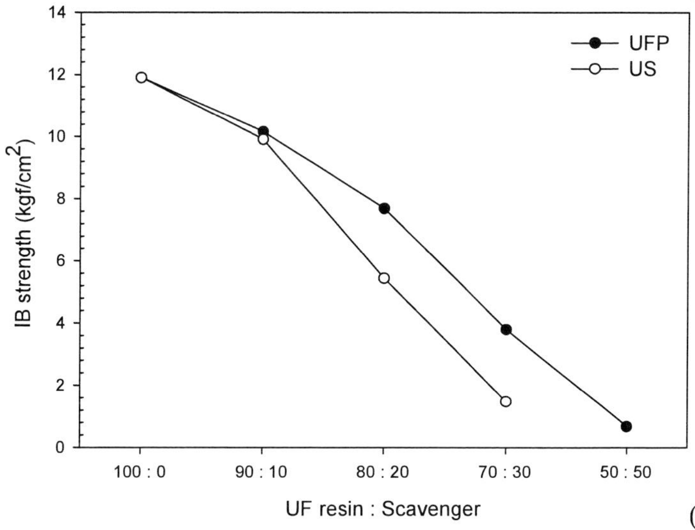

(b)

Figure 40. Properties of particleboards bonded with modified-UF resins with two scavengers. (a) Formaldehyde emission, and (b) IB strengths.

*Reproduced from ref. [66] with by permission from John Wiley & Sons (© John Wiley & Sons, 2008).

The Ea levels of the modified UF resins, depending on the concentration of either UFP or US, are shown in Figure 39, (a). When 10 wt% of US was added to the UF resin, the Ea of the resultant UF resin jumped from about 300 kJ/mole to about 542 kJ/mole, and then it gradually decreased to 358.7 kJ/mole (Figure 39, a). However, when the 10 wt% UFP was added to the UF resin, the Ea only slightly increased to 349 kJ/mole followed by a decrease to 258.2 kJ/mole. The Ea of the modified UF resins increased at a lower concentration and then it decreased as both US and UFP concentrations increased. In general, the Ea of the UFP-modified UF resin was much lower than those of the US-modified ones. This result suggests that the UFP-modified UF resins require less energy to start its curing process than that of the US-modified UF resins.

The Ea values of the UF resins that were prepared under the alkaline, weak acidic, and strong acidic conditions were about 78 kJ/mol, 94.8 kJ/mol, and 152.2 kJ/mol, respectively [60], which were greater than the reported one [37]. However, the Ea values are greater than these results. This could be attributed to many factors such as different F/U mole ratios, reaction conditions, and resin formulations. In spite of these inconsistencies, there is an inherent inaccuracy in the methods that are used to determine Ea values of a reaction system [61]. The author reported that the maximum rate method was more accurate than the dynamic method, which was used in this chapter.

Figure 39 (b) displays the changes of the heat of reaction (ΔH) of the UF resins modified with two scavengers. The ΔH is defined as the area under an exothermic peak of a DSC curve. The ΔH of the UFP-modified UF resin increased to 30% and then decreased. The US-modified UF resin also showed a similar trend, increasing to 20% and then decreasing. In general, the ΔH values of the UFP-modified UF resins were greater than those of the US-modified UF resins. Since the onset and peak temperatures increased with an increase in the US concentration in UF resin, it was expected that the US-modified UF resins would have greater ΔH values than those of the UFP-modified UF resin. Previous research has also reported an increase in the ΔH of UF resin when the peak temperature of the UF resin decreased with a decrease in the F/U mole ratio [40], but, this was not the case for the present study. The result in the current study could be attributed to the presence of methylolated ureas in the UFP. Methylolated ureas of the UFP would have facilitated to start the curing reaction of UF resin, which reduced the Ea of the UFP-modified UF resins, as shown in Figure 39 (a), but, an increase in the number of methylolated ureas provides more energy to form methylene or methylene ether linkages in order to complete the curing of UFP-modified UF resin. By contrast, it was thought that the urea grain dissolved in the US stayed in part as a separated substance which was not reacted with free formaldehyde that was insufficient in the US-modified UF resins. Further research on the chemical structure and species in modified UF resins is needed.

Properties (FE and IB strength) of particleboards bonded with either UFP- or US-modified UF resins are shown in Figure 40. When the UF resin was modified with the addition of 50% US, it was not able to prepare particleboard because of a de-lamination at the core layer. This could be due to high moisture content at the core layer by adding the US that composed of 40% urea and 60% water. As expected, the FE of particleboard continuously decreased with increasing the concentrations of two scavengers (Figure 40, a). In general, the FE values were smaller for particleboards bonded with the UFP-modified UF resin than those of the US-modified UF resins at all scavenger concentrations. This result suggests that the UFP is more effective in scavenging formaldehyde that the US does. However, the formaldehyde scavenging effectiveness of two scavengers should be judged by taking particleboard properties into account. Figure 40 (b) shows the IB strength values of particleboards bonded with either UFP- or US-modified UF resins, which is one of the critical properties of particleboard. The IB strength continuously decreased with an increase in the scavenger concentration regardless of the types of scavenger. But, the IB strength values of particleboard bonded with UFP-modified UF resins were greater than those of US-modified UF resins. And the IB strength of particleboard decreased from ~7.7 kgf/cm^2 to ~3.8 kgf/cm^2, which was an half when 20% of the UFP was added into the UF resin. This result could be attributed to a better reactivity of UFP-modified UF resin than that of US-modified UF resin, which was confirmed by onset temperature, peak temperature, or rate constants from the result of DSC. In other words, 20% addition of the UFP into UF resin reduced the FE of particleboard, and simultaneously did not cause a significant deterioration of the IB strength. This result indicates that the UFP is better in scavenging the FE of UF resin than the US does. When the FE and IB strength values of particleboard were taken into account, an optimum addition level of the UFP was determined as 20%, which resulted in a balance of lowering the FE and maintaining the IB strength at the same time.

Other properties of particleboard such as MOR), MOE, thickness swelling, and water absorption are presented in Table 11. MOR values did not change much at the 10% concentration of either the UFP, or US in UF resin, and then continuously decreased with an increase in the concentration of two scavengers. However, the MOR values of particleboard bonded with UFP-modified UF resin were much greater than those bonded with US-modified UF resin. These results could also be attributed to a better reactivity of the UFP-modified UF resin. MOE values of particleboards bonded with either UFP-, or US-modified UF resin slightly increased at the 10% concentration. As the scavenger concentration increased above 10%, MOE values gradually decreased for the UFP-modified UF resin, while it drastically decreased for US-modified UF resin. This result could be due to a greater densification of particles during hot-pressing, which resulted from more water sprayed when 10% scavenger was added into the UF resin.

The thickness swelling (TS) values of particleboard bonded with modified-UF resins continuously increased above 10% addition of two scavengers. And, the TS values were smaller for the particleboard bonded with the UFP-modified UF resin than those bonded with the US-modified UF resin, but they did not much change up to 20% concentration of two scavengers. The water absorption (WA) values of particleboards were also followed a similar trend to the TS value. These results could be attributed to a negative relation between the IB strength and TS, or WA. In other words, greater IB strength generally results in lower TS or WA for particleboard. A greater adhesive bond between particles could hold them together tight, which will result in a less TS and WA values in the particleboard. In addition, the TS and WA values of this chapter were much lower than those of the reported results [62]. These results indicated that the UFP was more effective than the US in scavenging the FE of UF resin adhesives and the addition of 20% UFP into UF resin was an optimum level when the properties of particleboard were taken into account.

5.2. Acrylamide Copolymerization of UF Resins

As shown in Table 12, the non-volatile solids contents of neat and modified UF resins were ranged from 55 to 64%. As expected, it increased as the acrylamide content increased. The resin viscosity slightly increased when the acrylamide increased up to 6% level. This result suggests that the molecular weights of the resins also follow a similar change. So, an increased viscosity of modified UF resins indicates a possible copolymerization of the acrylamide. The free formaldehyde content of modified UF resins was below the level of neat UF resin and 1% level gave the lowest value. These results suggest that the acrylamide reduces the free formaldehyde in the UF resin. The gel time of modified UF resins dramatically decreased when the acrylamide was added, but it slightly increased with an increase in the acrylamide content. These results indicate that the reactivity of UF resin is improved with added acrylamide and decreases with an increase in the acrylamide content.

In order to understand chemical structures of copolymerized UF resins, ^{13}C NMR spectroscopy was employed, and the results were shown in Figure 41. As shown in Figure 41 (a), two peaks at 125.7 and 131.7 ppm belonged to the pure acrylamide. And the peak at 89.5 ppm of control UF resin belonged to the methylene glycol, indicating the presence of un-reacted formaldehyde in the neat UF resin. After adding 1% acrylamide, two peaks at 89.5 and 131.7 ppm were disappeared (Figure 41, b). And it showed that the terminal amino group linked to carbonyl group via C=C bond of acrylamide reacted with either the methylene glycol or the methylol group. However, when the acrylamide increased to the 6% level, the peak at 131.7 ppm appeared again (Figure 41, c). It is believed this peak comes from the excess of the C=C linkages of acrylamide that are not reacted with the methylene glycol or the methylol groups.

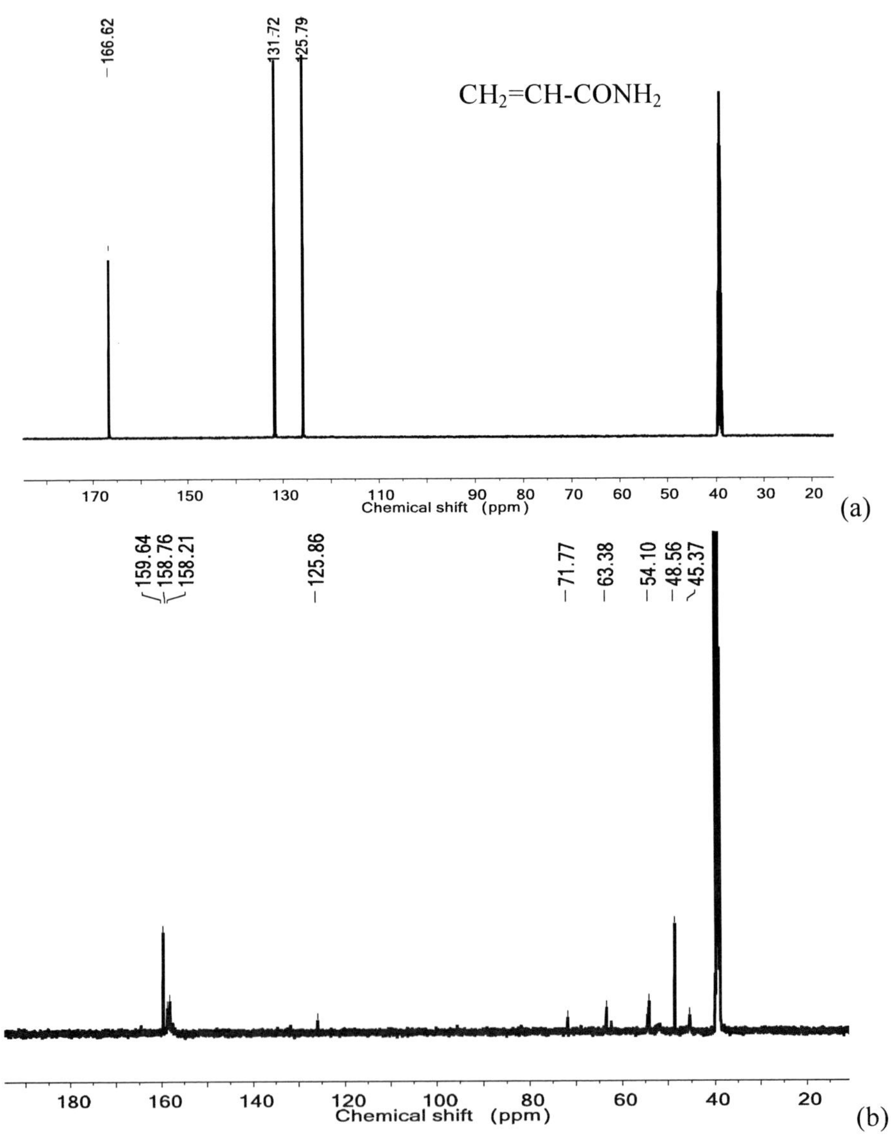

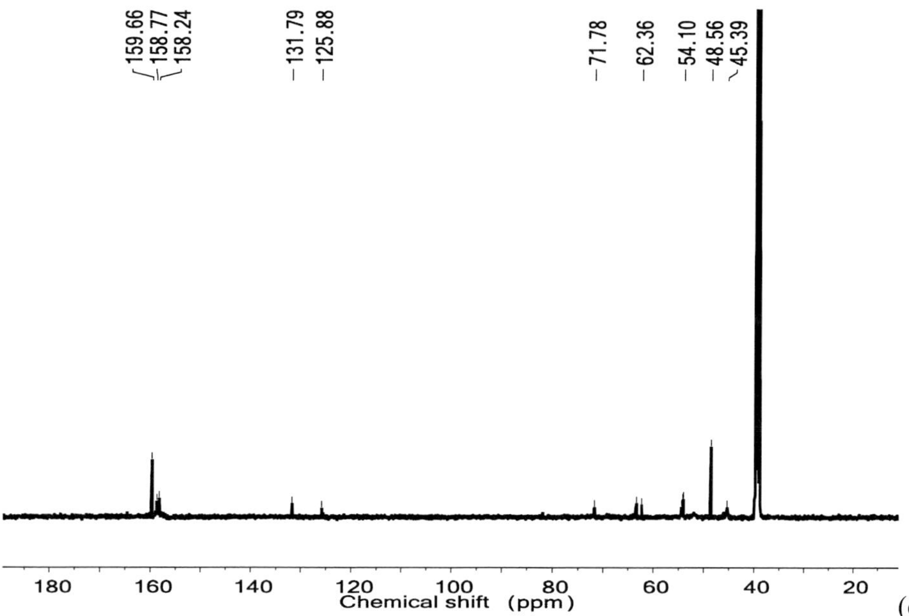

Figure 41. ^{13}C-NMR spectrum of modified UF resin. (a) Pure acrylamide, (b) 1% acrylamide, and (c) 6% acrylamide.

*Reproduced from ref. [68] with by permission from John Wiley & Sons (© John Wiley & Sons, 2010).

Figure 42 showed properties of plywood panels manufactured with copolymerized UF resin with acrylamide. All modified UF resins with different levels of the acrylamide release less FE than that of neat UF resin (Figure 42, a). This could be due to that the free formaldehyde content of the modified UF resin adhesive decreased with an increase of acrylamide level. Another reason might be that the acrylamide reacts with methylene glycol in UF resin adhesive during the copolymerization (Figure 41). The copolymerization of UF resin could reduce the number of terminal methylol groups that were supposed to be converted into methylene linkages by liberating formaldehyde during its curing [63]. Therefore, the acrylamide copolymerization lowers the FE of plywood bonded with the modified UF resin adhesives.

As a measure of adhesion performance, tensile shear strengths of plywood bonded with modified UF resin adhesives by different levels of acrylamide are given in Figure 42 (b). As the acrylamide content increased, the tensile shear strength initially increased and then decreased. The best adhesion strength was obtained for the modified UF resin with 1% acrylamide. Although there are many factors affecting the adhesion strength of plywood, some of the parameters that have contributed to an increased adhesion strength of plywood at 1% acrylamide addition could be a shorter gel time of the modified UF resin by adding 1% acrylamide (see Table 12). In other words, a faster gel time of the modified UF resin by adding 1% acrylamide produces more cross-links during the hot-pressing compared to those of the other modified UF resins, which eventually increases cohesive strength of the adhesive [64].

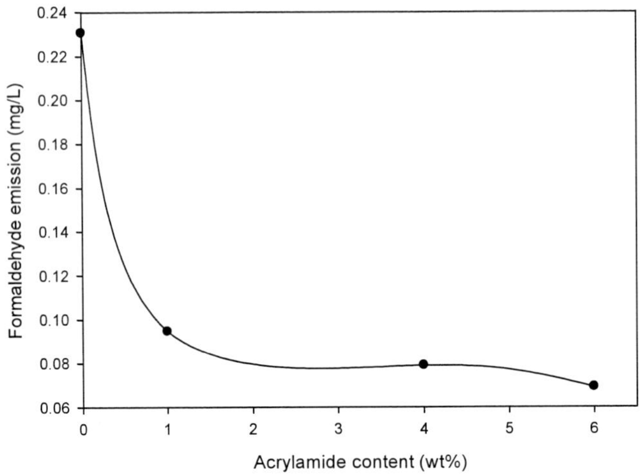

(a)

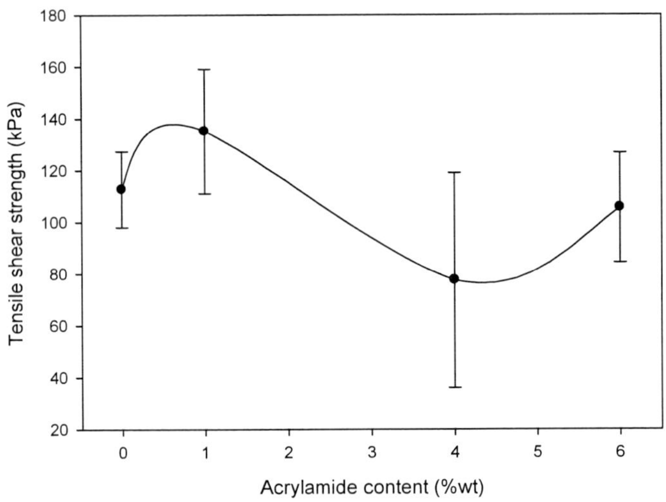

(b)

Figure 42. Properties of plywood's bonded with modified UF resins with different levels of acrylamide. (a) Formaldehyde emission, (b) tensile shear strength.
*Reproduced from ref. [68] with by permission from John Wiley & Sons (© John Wiley & Sons, 2010).

The average values of MOR and MOE of plywood, depending on different levels of acrylamide are given in Table 13. Compared to that of the control sample, MOR and MOE values of plywood bonded with the copolymerized-UF resin adhesives slightly decreased. This result could be ascribed to an increased viscosity of the copolymerized-UF resin adhesives as shown in Table 11. In other words, an increase in the viscosity of the modified UF resin adhesives might have influenced the adhesive penetration into the veneers during the manufacture of plywood. A slight penetration of the adhesives into the veneer could results in less stress transfer from veneer to the glue lines in plywood. However, it is believed that MOR and MOE of plywood might be affected by other factors as well, which is beyond the scope of this chapter. In general, it is believed that the addition of 1% acrylamide is an optimum level when the FE and tensile shear strength were taken into consideration.

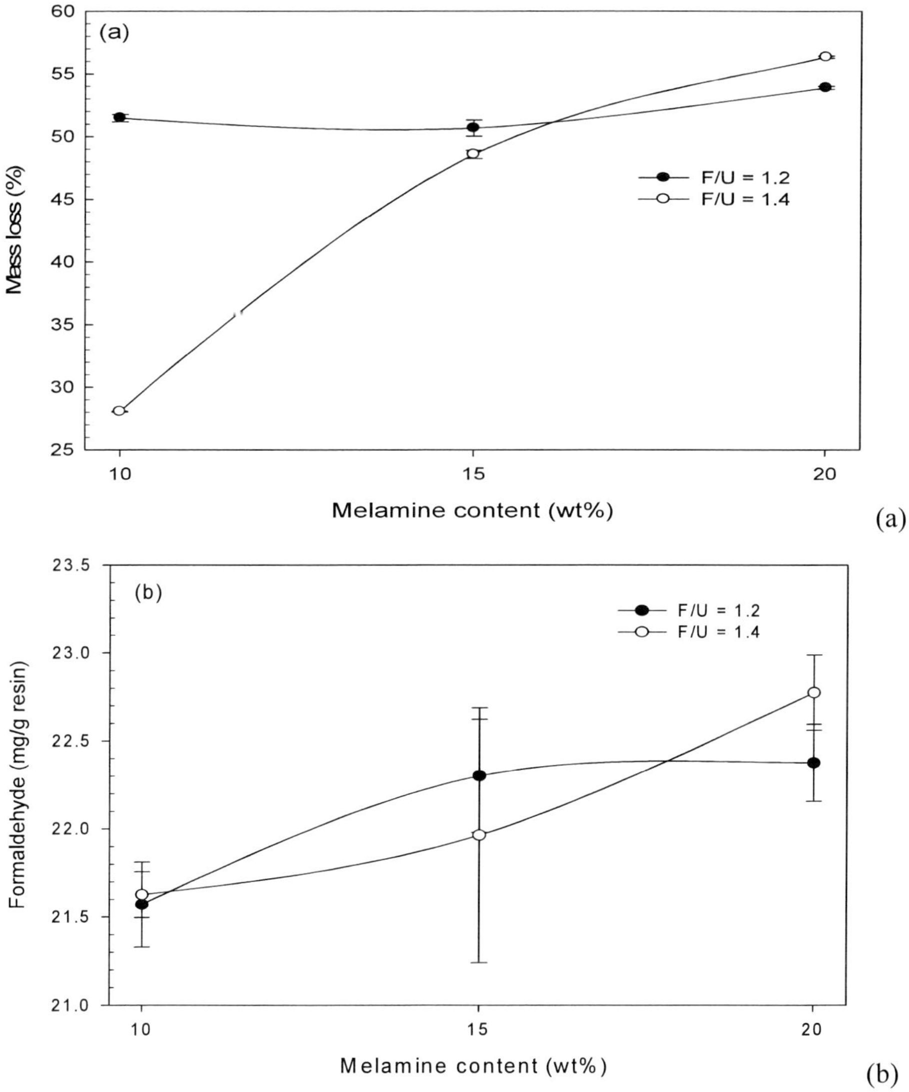

Figure 43. Hydrolytic stability of melamine-modified UF resins. (a) Mass loss, (b) liberated formaldehyde concentration.
*Reproduced from ref. [67] with by permission from John Wiley & Sons (© John Wiley & Sons, 2009).

Table 13. MOR and MOE values of plywood bonded with modified UF resins at different levels of acrylamide

Acrylamide level (%wt)	MOR (MPa)*	MOE (GPa)*
0	22.3±3.4	2.25±0.229
1	19.5±3.0	2.11±0.143
4	15.0±1.0	1.72±0.167
6	19.6±3.1	2.34±0.282

* MOR and MOE values show an average value with standard deviation.
Reproduced from ref. [68] with by permission from John Wiley & Sons (© John Wiley & Sons, 2010).

Table 14. Properties of the melamine –modified UF resins prepared at different F/U mole ratios

F/U mole ratio	Melamine content (%wt)	Gel time (s)	Non-volatile resin solids content (%)	Viscosity (mPa·s)	Free formaldehyde content (%)
1.2	10	128	57.8	108.0	0.93
	15	132	59.6	97.3	0.96
	20	119	60.8	136	0.97
1.4	10	119	60.9	176	1.01
	15	123	59.3	185.3	1.00
	20	113	61.0	238.7	0.95

Reproduced from ref. [67] with by permission from John Wiley & Sons (© John Wiley & Sons, 2009).

5.3. Melamine Modified UF Resins

As a part of understanding the hydrolytic degradation of UF resins, different amounts of melamine were added into UF resins to obtain melamine-modified UF (MUF) resins. And this section provides properties and hydrolytic stability of modified MUF resins. The properties of melamine–modified UF (MUF) resins prepared at different F/U mole ratios are summarized in Table 14. The non-volatile solids contents of MUF resins prepared at different F/U mole ratios were ranged from about 50 to 54 wt%. The resin viscosity and specific gravity was not much different for the F/U mole ratios.

As a measure of hydrolytic stability, the mass loss and liberated formaldehyde concentration of the MUF resins was also determined. Figure 43 shows changes of the mass loss and liberated formaldehyde concentration of the MUF resins, depending on two F/U mole ratios and three melamine contents. The mass loss of the cured MUF resin of a F/U mole ratio of 1.4 rapidly increased as the melamine content increased from 10% to 20% (Figure 43, a). However, the mass loss of the cured MUF resin with an F/U mole ratio of 1.2 did not change much, showing a slight increase at the 20% melamine content. These results suggested that the hydrolytic stability of cured MUF resins decreased as the melamine content and F/U mole ratio increased. Liberated formaldehyde concentrations of the MUF resins depending on the melamine content were presented in Figure 43 (b). In general, the liberated formaldehyde concentration of the resins increased with an increase in the melamine content. It was greater for the MUF resins with an F/U mole ratio of 1.4 at the melamine content of both 10% and 20%, except 15%. These results also indicated that hydrolytic stability, i.e. formaldehyde release of cured MUF resin decreased with an increase in both the melamine content and F/U mole ratio.

The results of mass loss and liberated formaldehyde concentration showed that the greater the melamine content and the higher the F/U mole ratio was the lower hydrolytic stability was. These results might be attributed to the chemical structure of MUF resins used. In other words, the greater melamine content and higher F/U mole ratio provided a more branched network structure with cured MUF resin. A more branched MUF resin has more pendant methylolgroups and a higher F/U mole ratio results in more ether linkages in UF resins [25]. These ether linkages are reversible reaction to produce methylolated ureas, which are also reversible to split into formaldehyde moieties in the end. It was reported that linear structures of UF resin had greater hydrolytic stability than branched structures of the UF resins [30]. These results showed that a higher F/U mole ratio and greater melamine content in the MUF resin decreased its formaldehyde release. But, it must be pointed out that the hydrolytic stability of cured MUF resin would quite differ from the one in wood-based composite panels such as plywood, particleboard, or fiberboard.

CONCLUSION

This chapter reported recent progresses on UF resins properties such as chemical structure, thermal curing behavior, hydrolytic stability, morphology, microstructure, crystalline structure, and resin modification as an adhesive for wood-based composite panels, particularly by focusing on synthesis reaction pH conditions, F/U mole ratio, and resin modifications in relation to its FE issue. Some of important conclusions were summarized as followings.

The amount of free formaldehyde strongly affected the reactivity of UF resin, and the molecular mobility of cured UF resin increased with decreasing the reaction pH used during its synthesis. These results indicated that the weak acid reaction condition provided a balance between increasing resin reactivity and improving adhesion strength of UF resins. UF resins with higher F/U mole ratios (i.e., 1.6 and 1.4) possess dimethylene ether linkages and methylene glycols, which give a greater contribution to the FE than that of lower F/U mole ratio. As the F/U mole ratio decreases, thermal curing behavior of these UF resins such as the gel time, onset and peak temperatures, and heat of reaction (ΔH) increased, while the activation energy (Ea) and rate constant (k) were decreased. The FE of particleboard (PB) was greatly reduced at the expense of the reactivity of UF resin and slight deterioration of performance of PB prepared when the F/U mole ratio decreased. DMA results partially explained the reason why UF resin with lower F/U mole ratio resulted in relatively poor adhesion performance.

Morphological investigation showed that the spherical structures in cured UF resins were much more resistant to the hydrolytic degradation by the acid than amorphous region. The soft phase detected by the AFM was much more susceptible to the hydrolytic degradation than the hard phase in cured UF resin. And the soft phase of cured UF resins by NH_4Cl was much more easily hydrolyzed than those cured by ammonium sulfate, indicating that hardener types had a great impact on the hydrolytic degradation behavior of cured UF resins. The presence of thin filament-like crystalline structures on the fracture surface of cured UF resin was reported. And the XRD results showed that the crystalline regions of cured UF resins with lower F/U mole ratio contribute partially to the improved hydrolytic stability of the cured resin.

Although there have been many progresses in understanding the properties of UF resins, further research work is still necessary to understand the properties of UF resins that eventually contribute to the formaldehyde emission in wood-based composites. For example, a better understanding on the mechanisms of hydrolytic degradation of UF resins is still needed to completely block or control the release of formaldehyde during the hydrolysis of cured UF resins. In addition, a detailed understanding of molecular architecture of cured UF resins will help design ways of blocking or controlling the hydrolytic degradation of UF resins. Future work is also needed to develop effective ways of modifying UF resins to improve hydrolytic stability of UF resins.

REFERENCES

[1] Johnson, R.S., An overview of North American wood adhesives resins, *In*: Wood Adhesives 2000, Forest Products Society, Madison, WI, 2001, pp:41~49.

[2] Lukkaroinen, J., & Dunky, M., European market for adhesives for panel board products: Actual state and challenges for the future, *In*: *Wood Adhesives 2005*, Forest Products Society, Madison, WI, 2006, pp: 39~44.

[3] Tomita, B. Wood Adhesive Trends in Asia, *In*: *Wood Adhesives 2005*, Forest Products Society, Madison, WI, 2006, pp: 45~50.

[4] Myers, G.E. (1983). Formaldehyde emission from particleboard and plywood paneling: measurement, mechanism, and product standards. *For. Prod. J.*, 33(5): 27~37.

[5] Dunky, M. (1998). Urea-formaldehyde adhesive resins for wood. *Int. J. Adhesion & Adhesives*. 18: 95-107.

[6] Myers, G.E. & Koutsky, J.A. (1987) Procedure for measuring formaldehyde liberation from formaldehyde-based resins. *For. Prod. J.* 37(9): 55~60.

[7] Myers, G.E. *In*: *Formaldehyde Release from Wood Products*, Ed. Meyer, B., Andrews, B.A.K., Reinhardt, R.M., American Chemical Society, 1986, pp: 8-14.

[8] Pizzi, A., Lipschitz, L. & Valenzuela, J. (1994). Theory and practices of the preparation of low formaldehyde emission UF adhesives. *Holzforschung*. 48: 254-261.

[9] Marutzky, R., *In*: *Wood Adhesives: Chemistry and Technology*, Vol. 2. Pizzi, A., Ed., Marcel Dekker Inc., 1986. pp: 307-387.

[10] Hse, C.Y., Xia, Z.Y. & Tomita, B. (1994). Effects of reaction pH on properties and performance of urea-formaldehyde resins. *Holzforschung*, 48(6): 527~534.

[11] Gu, J.Y., Higuchi, Morita, M. & Hse, C.Y. (1995). Synthetic conditions and chemical structures of urea-formaldehyde resins I. Properties of the resins synthesized three different procedures. *Mokkuzai Gakkaishi*, 41(12): 1115~1121.

[12] Steiner, P.R. (1973). Durability of urea-formaldehyde adhesives: Effects of molar ratio, second urea, and filler. *For. Prod. J.*, 23(12): 32~ 38.

[13] Myers, G.E. (1984). How mole ratio of UF resins affects formaldehyde emission and other properties: A literature critique, *For. Prod. J.*, 34(5): 35~41.

[14] Christjanson, P., Siimer, K., Pehk, T. & Lasn, I. (2002). Structural changes in urea-formaldehyde resins during storages, *Holz als Roh-und Werkstoff*, 60: 379-384.

[15] Pizzi, A., *Wood Adhesives: Chemistry and Technology*. Marcel Dekker Inc., New York, USA. 1983. pp: 59~104.

[16] Pizzi, A. *Advanced Wood Adhesives Technology*. Marcel Dekker Inc., New York. 1994. pp: 19-66.

[17] Soulard, C., Kamoun, C. & Pizzi, A. (1999). Uron and uron–urea-formaldehyde resins. *J. Appl. Polym. Sci.*, 72, 277~289.

[18] Tohmura, S., Hse, C.Y. & Higuchi, M. (2000). Formaldehyde emission and high-temperature stability of cured urea-formaldehyde resins. *J. Wood Sci.*, 46: 303~309.

[19] Tomita, B. & Hatono, S. (1978). Urea-formaldehyde resins. III. Constitutional characterization by ^{13}C fourier transform NMR spectroscopy. *J. Appl. Polym. Sci.*, 16: 2509~2525.

[20] Ebdon, J.R. & Heaton, P.E. (1977), Characterization of urea-formaldehyde adducts and resins by ^{13}C-n.m.r. spectroscopy, *Polymer*, 18: 971~974.

[21] Meyer, C.B., Formaldehyde release from urea-formaldehyde systems, Ed. Maloney, T.M., *In*: Proc. of the 13th *International Symposium on Particleboard*, Washington State University, Pullman, WS, USA, 1979, pp: 43~354.

[22] Kim, M.G. & Amos, L.W. (1990). Quantitative carbon-13 NMR study of urea-formaldehyde resins in relation to the formaldehyde emission levels, *Ind. Eng. Chem.*, 29: 208~212.

[23] Kim, M.G. (1999). Examination of selected synthesis parameters for typical wood adhesive-type urea-formaldehyde resins by ^{13}C NMR spectroscopy I, *J. Polym. Sci., Part A: Polym. Chem.*, 37: 995~1007.

[24] Kim, M.G. (2000). Examination of selected synthesis parameters for typical wood adhesive-type urea-formaldehyde resins by ^{13}C NMR spectroscopy. II, *J. Appl. Polym. Sci.*, 75: 1243~1254.

[25] Kim, M.G. (2001). Examination of selected synthesis parameters for typical wood adhesive-type urea-formaldehyde resins by ^{13}C NMR spectroscopy III, *J. Appl. Polym. Sci.*, 80: 2800~2814.

[26] Kim, M.G., Wan, H., No, B.Y. & Nieh, W.L. (2001). Examination of selected synthesis parameters for typical wood adhesive-type urea-formaldehyde resins by ^{13}C NMR spectroscopy IV, *J. Appl. Polym. Sci.*, 82: 1155~1169.

[27] Kim, M.G., Wan, H., No, B.Y., Lee, S.M. & Nieh, W.L. (2003). Examination of selected synthesis parameters for typical wood adhesive-type urea-formaldehyde resins by ^{13}C NMR spectroscopy IV, *J. Appl. Polym. Sci.*, 89: 1896~1917.

[28] Ferg, E.F., Pizzi, A. & Levendis, D.C. (1993). ^{13}C NMR analysis method for urea-formaldehyde resin strength and formaldehyde emission, *J. Appl. Polym. Sci.*, 50: 907~915.

[29] Pascault, J.-P., Sautereau, H., Verdu, J.R., Williams, J.J., *Thermosetting polymers*, Marcel Dekker Inc., New York, USA, 2002.

[30] Chung, I. & Maciel, G.E. (1994). NMR study of the stabilities of urea-formaldehyde resin components toward hydrolytic treatments. *J. Appl. Polym. Sci.*, 52: 1637-1651.

[31] Gu, J.Y., Higuchi, M., Morita, M. & Hse, C.Y. (1996). Synthetic conditions and chemical structures of urea-formaldehyde resins II. Synthetic procedures involving a condensation step under strongly acidic conditions and the properties of the resins obtained. *Mokkuzai Gakkaishi*, 42(2): 149~156.

[32] Soulard, C., Kamoun, C. & Pizzi, A. (1998). Uron and uron-urea-formaldehyde interior wood adhesives, *Holzforschung Holzverwertung*, 50(5): 89~94.

[33] Myers, G.E. (1981). Investigation of urea-formaldehyde polymer cure by infrared. *J. Appl. Polym. Sci.*, 26: 747-764.

[34] Socrates, G. Infrared Characteristic Group Frequencies. 2nd Edition, John Wiley & Sons, New York, 1994.

[35] Maciel, G.E., Szeverenyi, N.M., Early, T.A. & Myers, G.E. (1983). Carbon-13 NMR studies of solid urea-formaldehyde resins using cross polarization and magic-angle spinning. *Macromolecules*, 16: 598~604.

[36] Park, B.D., Kim, Y.S., Sing, A.P. & Lim, K.P. (2003). Reactivity, chemical structure, and molecular mobility of urea-formaldehyde (UF) adhesives synthesized under different conditions using FT-IR and solid state ^{13}C CP/MAS NMR spectroscopy. *J. Appl. Polym. Sci.*, 88(11): 2677-2687.

[37] Pizzi, A., & Panamgama, L.A. (1995). Diffusion hindrance vs. wood-induced catalytic activation of MUF adhesive polycondensation, *J. Appl. Polym. Sci.*, 58: 109~115.

[38] Siimer, K., Kaljuvee, T. & Christjanson, P. (2003). Thermal behaviour of urea-formaldehyde resins during curing. *J. Therm. Anal. Calorim.*, 72: 607~617.

[39] No, B. & Kim, M.G. (2005). Curing of low level melamine-modified urea-formaldehyde particleboard binder resins studied with dynamic mechanical analysis (DMA). *J. Appl. Polym. Sci*. 97: 377~389.

[40] Park, B.D., Kang, E.C. & Park, J.Y. (2006a). Effects of formaldehyde to urea mole ratio on thermal curing behavior of urea-formaldehyde resin and properties of particleboard, *J. Appl. Polym. Sci.*, 101(3): 1787~1792.

[41] Onic, L., Bucur, V., Ansell, M.P., Pizzi, A., Deglise, X. & Merlin, A. (1988). Dynamic thermomechanical analysis as a control technique for thermoset bonding of wood joints. *Int. J. Adhesion & Adhesives*, 18: 89~94.

[42] Kim, M.G., Nieh W.L.S. & Meacham, R.M. (1991). Study on the curing of phenol-formaldehyde resol resins by dynamic mechanical analysis. *Ind. Eng. Chem. Res*. 30: 798~803.

[43] Flory, P. J., *Principles of Polymer Chemistry*, Cornell University Press, Ithaca, NY, US, 1953.

[44] Ward, I.M. *Mechanical Properties of Solid polymers*, Wiley-Interscience, London, 1971.

[45] Cullity, B.D. & Stock, S.R. Elements of X-ray Diffraction, 2001, Prentice-Hall Inc., New Jersey, USA. pp: 167~184.

[46] Despres, A. & Pizzi, A. (2006). Colloidal aggregation of aminoplastic polycondensation resins: urea-formaldehyde versus melamine-formaldehyde and melamine-urea-formaldehyde resins. *J. Appl. Polym. Sci*. 100: 1406~1412.

[47] Stuligross, J. & Kousky, J.A. (1985). A morphological study of urea-formaldehyde resins, *J. Adhes*. 18: 281~299.

[48] Johns, W.E., & Dunker, A.K. (1986) Urea-formaldehyde resins. In: *Formaldehyde Release from Wood Products*, Ed., Meyer B, Andrews B.A.K., Reinhardt, R.M., American Chemical Society. pp: 76~86.

[49] Celzard, A. Pizzi, A., & Fifero, V. (2008). Physical gelation of water-borne thermosetting resins by percolation theory-urea-formaldehyde, melamine-urea-formaldehyde, and melamine-formaldehyde resins. *J. Polym. Sci.: Part B: Polym. Phys*. 46: 971~978.

[50] Elbert, A.A. (1995). Influence of hardener systems and wood on the formaldehyde emission from urea-formaldehyde resin and particleboards, *Holzforschung*, 49: 358~362.

[51] Newbury, D.E. *Advanced Scanning Electron Microscopy and X-ray Microanalysis*, 1986, New York, Plenum Press, USA.

[52] Pratt, T.J., Johns, W.E., Rammon, R.M., & Plagemann, W.L. (1985). A novel concept on the structure of cured urea-formaldehyde resin. *J. Adhesion*. 17: 275~295.

[53] Vanlandingham, M.R., Eduljee, R.F. & Gillespie Jr., J.W. (1999). Relationships between stoichiometry, microstructure, and properties for amine-cured epoxies. *J. Appl. Polym. Sci.*, 71(5): 699~712.

[54] Park, B.D. & Jeong, H.W. (2011a). Effects of acid hydrolysis on microstructure of cured urea-formaldehyde resins using atomic force microscopy, *J. Appl. Polym. Sci.*, 122: 3255-3262.

[55] Motter, W.K. (1990). The Formation of the Colloidal phase in Low Mole Ratio Urea-Formaldehyde Resins. Ph.D. Thesis, Department of Mechanical and Materials Engineering, Washington State University, Pullman, WA, USA.

[56] Levendis, D., Pizzi, A. & Ferg, E. (1992). The correlation of strength and formaldehyde emission with crystalline/amorphous structure of UF resins. *Holzforschung*, 46(3): 263~269.

[57] Dunker, A.K., Johns, W.E., Rammon, R., Framer, B. & Johns. S. (1986). Slightly bizarre protein chemistry: urea-formaldehyde resin from a biochemical perspective. *J Adhesion*, 19: 153~176.

[58] Park, B.D., Jeong, H.W. & Lee, S.M., (2011). Morphology and chemical elements detection of cured urea-formaldehyde resins. *J. Appl. Polym. Sci.*, 120; 1475-1482.

[59] Park, B.D. & Jeong, H.W. (2011b). Hydrolytic stability and crystallinity of cured urea-formaldehyde resins with different formaldehyde/urea mole ratio, *Int. J. Adhesion & Adhe*sives, 31: 524-529.

[60] Park, B.D., Kang, E.C. & Park, Y.J. (2006b). Differential scanning calorimetry of urea-formaldehyde adhesive resins, synthesized under different pH conditions. *J. Appl. Polym. Sci.*, 100(1): 422-427.

[61] Starink, M.J. (2003). The determination of activation energy from linear heating rate experiments: a comparison of the accuracy of isoconversion methods. *Thermochimica Acta*, 404: 163~176.

[62] Nieh, W.L.S. *In: 1998 Resin & Blending Seminar Proceedings,* Ed. Bradfield, J., The Composite Panel Association, Gaithersburg, MD, 1999, pp: 23-27.

[63] Myers, G.E. (1982). Hydrolytic stability of cured urea-formaldehyde resins, *Wood Sci., 15(2): 127-138.*

[64] Bolton, A.J. & Irle, M.A. (1987). Physical aspects of wood adhesive bond formation with formaldehyde based adhesives Part I. The effect of curing conditions on the physical properties of urea formaldehyde films, *Holzforchung*, 41: 155~158.

[65] Park, B.D. & Kim, J.W. (2008). Dynamic mechanical analysis of urea-formaldehyde resin adhesives with different formaldehyde to urea mole ratios, *J. Appl. Polym. Sci.*, 108(3): 2045-2051.

[66] Park, B.D. Kang, E.C. & Park, J.Y. (2008). Thermal curing behavior of modified urea-formaldehyde resins adhesives with two formaldehyde scavengers and their influences to adhesion performance. *J. Appl. Polym. Sci*. 110(3): 1573~ 1580.

[67] Park, B.D., Lee, S.M. & Roh, J.K. (2009). Effects of formaldehyde/urea mole ratio and melamine content on the hydrolytic stability of cured urea-melamine-formaldehyde resin. *Holz als Roh- und Werkstoff*, 67(1): 121~123.

[68] Abdullah, Z.A. & Park, B.D. (2010). Influence of acrylamide copolymerization of urea-formaldehyde resin adhesives to their chemical structure and performance, *J. Appl. Polym. Sci.*, 117(6): 3181~3186.

[69] Park, B.D., Lee, S.M. & Park, J.Y. (2008). ^{13}C-NMR spectroscopy of urea-formaldehyde resin adhesives with different formaldehyde/urea mole ratios, *Mokchae Konghak*, 36(2): 63~72.

[70] Park, B.D. & Jeong, H.W. (2011). Acid hydrolysis influence on the morphology of cured urea-formaldehyde resins of different formaldehyde/urea mole ratios, *Mokchae Konghak*, 39(2): 179-186.

In: Formaldehyde
Editors: Chan Bao Cheng and Feng Hu Ln

ISBN 978-1-62257-214-4
©2012 Nova Science Publishers, Inc.

Chapter 2

FORMALDEHYDE EMISSIONS FROM WOOD-BASED PANELS: TESTING METHODS AND INDUSTRIAL PERSPECTIVES

Luisa H. Carvalho*
DEMad - Departamento de Engenharia de Madeiras,
Instituto Politécnico de Viseu, Campus Politécnico de Repeses
Viseu, Portugal
Fernão D. Magalhães†
LEPAE - Departamento de Engenharia Química,
Faculdade de Engenharia da Universidade do Porto, Porto, Portugal
João M. Ferra‡
Euroresinas – Industrias Químicas, S. A., Sines, Portugal

ABSTRACT

Formaldehyde is an important chemical feedstock for the production of phenoplast and aminoplast thermosetting resins, by reaction with other monomers (mostly urea, but also melamine, phenol and resorcinol). These adhesives are mainly used in the manufacture of wood-based panels: plywood, particleboard, hardboard, medium density fiberboard (MDF) and oriented strand board (OSB). These products have a wide range of applications, from non-structural to structural, outdoor or indoor, mostly in construction and furniture, but also in decoration and packaging. The WBP industry plays an important role in the global economy and contributes for forest sustainability and carbon sequestration. In 2009, FAO (Food and Agriculture Organization) reported that a total of 260 million m^3 WBPs were produced in the world (Europe 29.7%, Asia 43.9%, North America 18.3% and others 2.5%).

* E-mail: lhcarvalho@demad.estv.ipv.pt.
† E-mail: fdmagalh@fe.up.pt.
‡ E-mail: joao.ferra@sonaeindustria.com.

Being economically competitive and highly performing, a major drawback of formaldehyde-based resins, mostly urea-formaldehyde, is the formaldehyde emission during panel manufacturing and service life. There are two sources of emission: release of unreacted monomer, during or after panel production, and long-term resin degradation (hydrolysis). The formaldehyde content and chemical stability of the resin will therefore affect emission levels. In addition, external factors like temperature, humidity or air renewal rate will also play a role. It must be noted that wood itself contributes to formaldehyde emission, since it is a product of metabolism and decomposition processes. The actual emission level depends strongly on the type(s) of wood used in panel production.

Due to information considering formaldehyde as potentially carcinogenic to humans, the implementation of international regulations and requirements for emissions from WBPs has led to establishment of standard testing methods. Two main groups are considered: chamber methods (emulating indoor living environments, mentioned in ASTM, ISO and European standards), and small scale methods, also called derived tests, oriented to industrial quality control and development. This second group includes commonly used methods, mentioned in different international standards, like the so-called: perforator (actually a test of potential formaldehyde emission), flask, desiccator, and gas analysis methods. Correlation between results from different methods has been a matter of debate, not yet completely elucidated.

Based on different test methods, emission limit standards for WBPs have been issued by several governmental organizations in Europe, Japan and United States, allowing for product classification according to emission level. Additionally, limits drawn by major industrial consumers, like IKEA, have been a defining guideline for WBP producers.

In order to comply with increasingly stringent requirements, the industry has been developing strategies to minimize formaldehyde emissions from WBPs. Four major approaches can be found: 1) reduction of formaldehyde content in resin formulation, while attempting to maintain adhesive performance, 2) addition of formaldehyde scavengers to resin or wood particles, having the negative effect of consuming formaldehyde prior to resin cure, 3) implementation of surface treatments after board production, and 4) use of alternative adhesive systems with reduced or no emissions, with an impact on product cost and/or performance.

1. Industrial Production of Wood-based Panels

Wood-based panels (WBPs) are manufactured from wood materials having various geometries (e.g., fibers, particles, strands, flakes, veneers, and lumber), combined with an adhesive, and bonded in a press. The press applies heat (if needed) and pressure to activate (chemically cross-link) the adhesive resin and bond the wood material into a solid panel having good mechanical and physical properties (strength, stiffness, form, dimensional stability, etc.)

The most used wood-based panels are plywood, particleboard (PB), medium density fiberboard (MDF) and oriented strand board (OSB). Other examples of wood-based panels are hardboard, LVL-laminated veneer lumber, SWP-solid wood panels and cement-bonded particleboard. Modern plywood, made by gluing together several hardwood veneers or plies, was the first type of wood-based panel produced in 1935 in Portland, USA (APA). Only 60 years later particleboard panels were manufactured. Figure 2.1 summarizes the classification of wood-based panels according to particle size, density, and process type (Suchsland and Woodson, 1987).

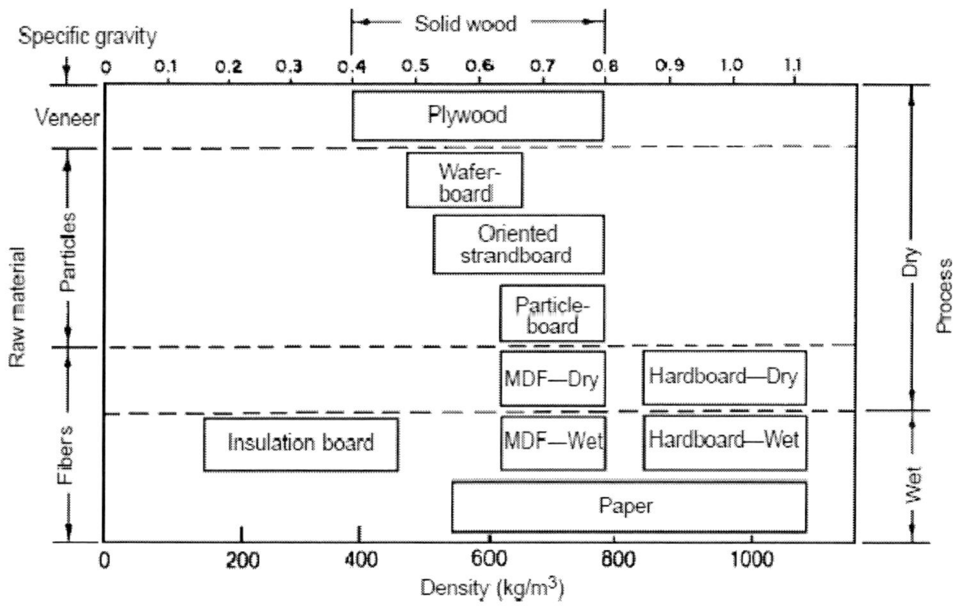

Source: (Suchsland and Woodson, 1987).

Figure 2.1. Classification of wood-based panels by particle size, density, and process type.

1.1. Manufacture of Particleboard

Particleboard is manufactured from wood chips, sawdust, waste materials and recycled woodchips (Youngquist, 1999). Typically, it is made in three layers. The two external layers consist of finer particles and sawdust, while the core layer is made of coarser material.

The manufacture of particleboard has five main steps: (1) furnish preparation, (2) resin application, (3) mat formation, (4) hot pressing, and (5) finishing. The furnish is prepared by refining the raw materials into small particles and drying them to achieve a desired moisture content, about 2 to 7 % (Youngquist, 1999). The type of resin used in particleboard depends of the characteristics desired, but normally urea-formaldehyde (UF) resin is used. The resin/wood ratio, based on resin dry solids content, and particle dry weight, is usually 6 to 9 % (Youngquist, 1999; Dunky, 2003). Additives to enhance characteristics like fire retardancy or moisture resistance can be applied at this stage. After mechanically mixing the particles and the adhesive system, the material goes through a continuous mat-forming system and is then hot-pressed under pressures between two and three MPa and temperatures between 140 °C and 220 °C (Youngquist, 1999; Dunky, 2003). After the press cycle is complete, the panel is transported to a board cooler, and then hot-stacked to wait sawing into finished panel sizes and sanding.

1.2. Manufacture of OSB

OSB (oriented strand board) is a structural building material used for residential and commercial construction. It is a multi-layered board mainly made from strands of wood together with a binder. The strands in the external layer are aligned and parallel to the board length or width. The strands in the internal layer or layers can be randomly orientated or aligned, generally at right angles to the strands in the external layers (EN 300). The manufacture process of OSB is very similar to particleboard. The main differences are the type of particles, resination process and mat formation (Marra, 1992).

Typically OSB is made from freshly harvested aspen poplar, pine or other mixed hardwood and softwood logs, recovered wood are not use. Phenol-formaldehyde (PF) resin and pMDI are the most commonly adhesives used in OSB industries, although melamine urea formaldehyde (MUF) resins and melamine and/or urea modified PF resins are also used to decrease the price of adhesive (Dunky, 2003, Marra, 1992). Irle and Barbu (2010) reported that the current trend in Europe is the use of pMDI adhesive on face (3-6 %) and core (4-10 %) due to the low formaldehyde emission and short pressing time.

1.3. Manufacture of MDF

MDF (medium density fiber) panels consist of ligno-cellulosic fibres manufactured by the "dry process", i.e. having a fiber moisture content less than 20 % at the forming stage and being essentially produced under heat and pressure, after mixing with adhesive and wax. Fibers are usually obtained from a thermo mechanical pulping process, which consists in the combination of heat and mechanical energy to break the bonds between wood cells (Irle and Barbu, 2010). UF and fortified UF resins are the most used adhesives to manufacture MDF. In specifics products, requiring moisture resistance and fire retardancy, MUF resins and pMDI are used. The resins are sprayed onto wood fibers passing in a blowline. Understanding and optimizing this step is the most challenging task in the MDF process (Waters, 1990). According to Chapman (Chapman, 2011) the optimization of the blowline and the resin injection nozzles permits to reduce significantly resin consumption.

1.4. Manufacture of Plywood

Plywood is a composite panel made from thin layers of wood veneer. The layers are glued together under heat and pressure, each with its grain at right angles to adjacent layers to improve strength (Sellers, 1985). Usually UF resins are used to produce interior boards without special requirements concerning water resistance. PF and MUF or MUPF resins are used for making exterior plywood (Dunky, 2003). The plywood manufacturing process has three main stages: 1) log preparation, 2) veneer plain slicing or rotary cutting, drying and grading, and 3) board lay-up, pressing and finishing (Irle and Barbu, 2010). Resin grammage typically ranges from 140 to 240 g/m^2 per face that depend of the type of wood veneer and the operation conditions, temperature and pressing times (Irle and Barbu, 2010).

Plywood is considered a material of choice in the building industry because of outstanding structural performance, as defined by a high strength-to-weight ratio, excellent dimensional stability, and durability compared to other building material. Due to the high price, plywood has been substituted by OSB in specific applications.

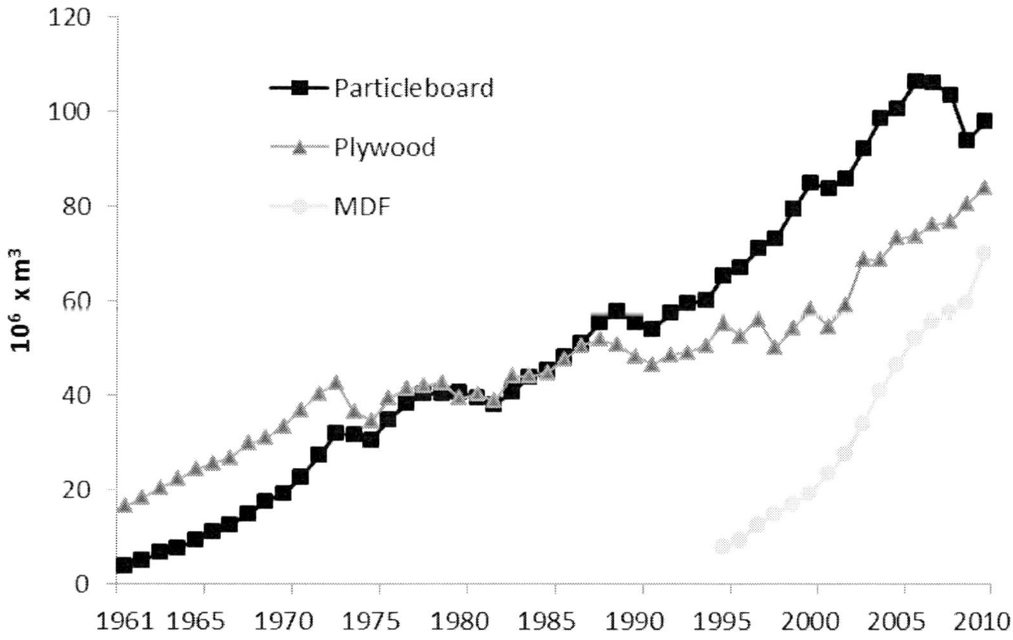

Figure 2.2. Evolution of the global production of wood-based panels. Source: (FAO, 2011).

1.5. Market

The increase of the world demand for wood-based composites and the awareness of the tree role in the global ecosystem are driving the use of recycled wood and wood from different sources/species in the formulation of wood composites (Carvalho, 1999). The variability of available wood creates difficulties concerning the compatibility/adequacy of the resin (binding agent) with the wood (Sigvartsen and Dunky, 2005).

Food and Agriculture Organization of the United Nations (FAO, 2011) reported that in 2010 approximately 100 million m^3 of particleboard (EUA, Germany, Canada and China manufacture 20 %, 10 %, 9 % and 8 % respectively), 70 million m^3 of MDF (China, Germany and EUA manufacture 45 %, 8 % and 6 % respectively) and 84 million m^3 of plywood (EUA, Germany, Canada and China manufacture 20 %, 10 %, 9 % and 8 % respectively) were manufactured in the world (see Figure 2.2) (FAO, 2011).

European Panel Federation (EPF) reported that the wood-based panels industry was affected by the economic crisis in 2008 (Wijnendaele, 2009), in particular the production of particleboard and MDF, which decreased in 2008 by 8.7 % and 8 % respectively (Wijnendaele, 2009).

1.6. Environmental Impact

The European woodworking industry stands for about 100,000 companies, two million employees and an annual turnover of 150 billion € (EPF, 2011). Furthermore, forests and forest-based industries provide direct employment to three million people throughout the EU, especially in remote areas (EPF, 2011). They represent 10 % of the total production value of the EU manufacturing industry (EPF, 2011). According to European woodworking industry, these businesses invest continuously in sustainable forest management, deflorestation and reforestation activities to ensure reliable wood availability.

Wood is formed by photosynthesis of CO_2 and water, thereby blocking carbon in a durable way. During growth a tree absorbs, through photosynthesis, approximately the equivalent of 1 ton of CO_2 for every m^3 growth, while producing the equivalent of 0.7 ton of oxygen (CEI-Bois, 2007). Wood therefore plays a major role in fighting climate changes. Rational use of wood sources stimulates forest expansion and reduces greenhouse gas emissions.

Wood products require less energy for manufacturing (up to 6000 MJ/m^3) than alternative raw materials, hence contributing even more to the reduction of fossil fuel consumption. By using the full potential of wood (sink and substitution effects) in buildings, Europe could reduce emissions of CO_2 with 300 million ton or 15 to 20% (EPF, 2011).

The recycling process has a great paper in future of wood-based panels industry. In 2004 the proportion of recycled wood used in manufacturing of particleboard was 23 % (EPF, 2005). Roffael et al. (2009) studied the use of recycled fiberboards with raw material to making MDF. They concluded that the use of waste fiberboards up to 33 % does not have effect on the mechanical properties of the panels.

In 2003 the European Woodworking Industries, Pulp and Paper Industries and the European Commission created a work group for discuss the use of the wood sources with energy and wood products (CEI-Bois, 2007). The main recommendation was to consider "wood-based products as carbon sinks under the Kyoto Protocol, thereby acknowledging the contribution of wood-based products to climate change mitigation and the carbon cycle, and recognize their superior eco-efficiency versus other materials, as well as their outstanding properties in recycling with minimal energy use" (CEI-Bois, 2007).

2. Current Use of Formaldehyde-based Resins

Formaldehyde is an important chemical for the global economy, widely used in the production of thermosetting resins, as an intermediate raw material in the synthesis of several chemicals, and for preservation and disinfection (Global Insight Inc., 2006; Tang et al., 2009). The annual world production is about 21 million ton. Figure 3.1 summarizes the industrial uses of formaldehyde and related products. Production of urea-formaldehyde, phenol-formaldehyde, and melamine-formaldehyde resins accounts for about 50 % of global formaldehyde consumption (Global Insight Inc., 2006). In 2003, the value of sales of formaldehyde and derivative products in United States and Canada reached approximately USD$ 145 billion. The number of workers involved in related activities was reportedly 4.2 million, which represents nearly 3.4 % of employment in private, nonfarm establishments in North America (Global Insight Inc., 2006).

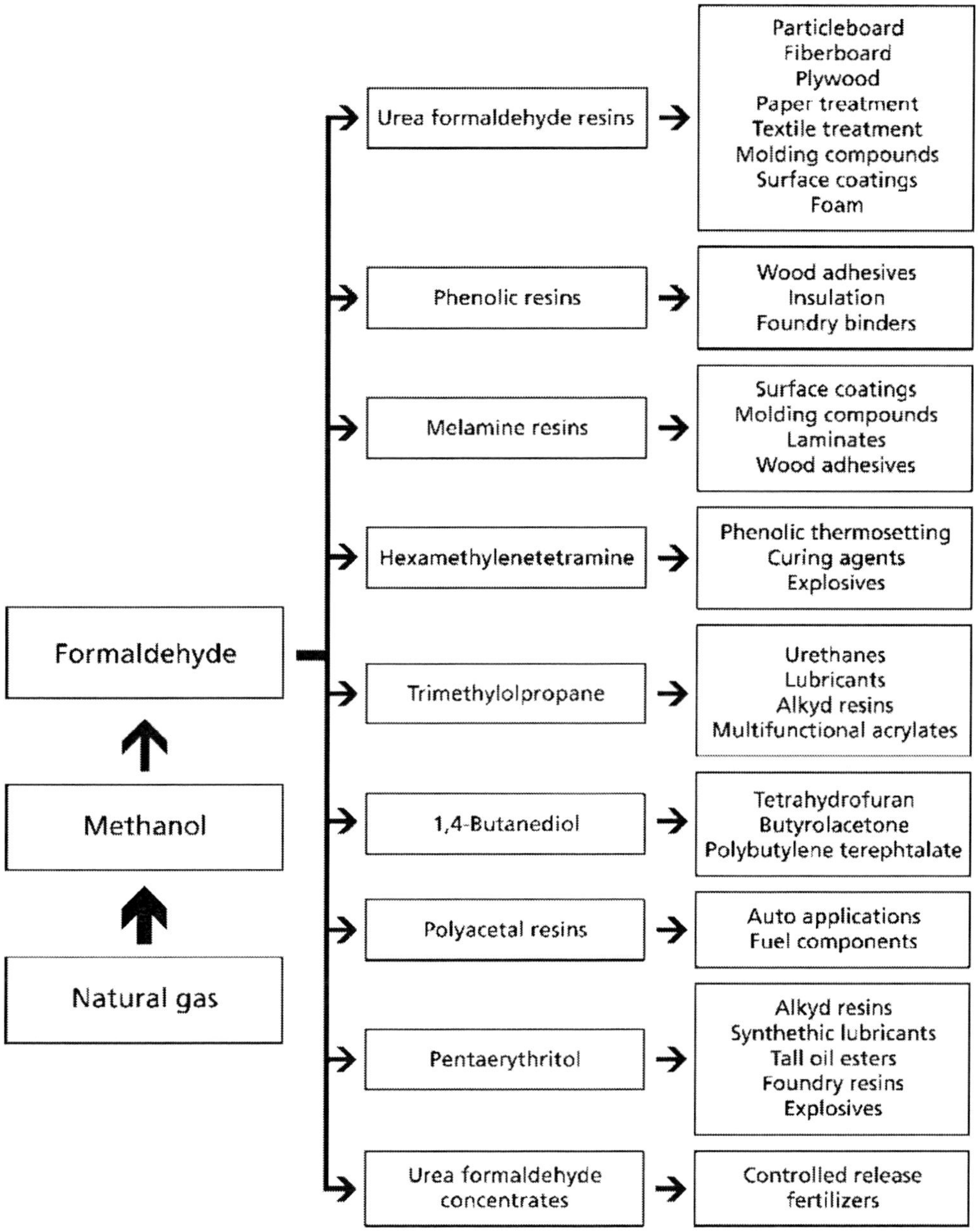

Figure 3.1. Product Tree for Formaldehyde. Reprinted with permission from: Salthammer, T., Mentese, S., Marutzky, R., "Formaldehyde in the indoor environment" Chemical reviews, 110, 2536-72, 2010. Copyright 2010 American Chemical Society.

2.1. Urea-formaldehyde Resins

Urea-formaldehyde (UF) polymers have been for decades the most widely used adhesives in the manufacture of wood-based panels, such as particleboard (PB), medium density fiberboard (MDF) (both consuming 68 % of the world's UF resins production) and plywood (consuming 23 %) (SRI, 2009).

According to SRI Consulting (SRI, 2009) the global production of UF resins in 2008 was approximately 14 million ton. Their consumption increased 2.8 % in 2008, and is expected to grow an average 3.2 % per year from 2008 to 2013, and just under 2 % per year from 2013 to 2018. Table 3.1 describes the main uses of UF resins (Dunky, 1998; Dunky and Pizzi, 2002; Dunky, 2003).

Table 3.1. Main applications of urea-formaldehyde resins

Application	%
Wood Composites: adhesives for non-structural panels, particleboard and MDF, for structural panels, plywood and OSB.	95
Cork Composites: adhesive for interior application of cork panels	5
Decorative e protective laminates: countertops, cabinets, furniture, flooring, wall covering, sheathing, automobile interiors	
Paper treating and coating: wet-strength resins added to sanitary paper products such as facial tissue, table napkins, and roll towels.	
Surface coatings: crosslinkers in stove paint systems in combination with other polymeric materials including alkyds, acrylics, epoxies and saturated polyesters.	
Textile treatment: printing inks, dyes and textile finishing products (crease-resistant textile products e.g. products that does not wrinkle easily).	
Foundry materials and binders (adhesives in molds to produce castings): sand binder to coat sand, which is then used in core making for casting operations in the foundry industry.	
Fiberglass and rock wool insulation: specific applications include low-density insulation, high-density industrial insulation, and other specialty insulation.	
Molded plastic products: electrical switches, circuit breakers, stove hardware, buttons and housings.	
Abrasive materials: coated and bonded abrasives.	

The main reasons for the wide use of UF resin in wood based panels are high reactivity, low cost and excellent adhesion to wood. One the other hand, the most important drawbacks are low moisture resistance and formaldehyde emission during panel manufacture and service life (Pizzi, 2003; Dunky, 1996). Although free formaldehyde content on these resins has been decreased during the last decades, the recent reclassification of formaldehyde by International Agency for Research on Cancer (IARC) as "carcinogenic to humans", is forcing resin producers to develop systems that lead to a decrease in its emissions to levels as low as the present in natural wood (Athanassiadou, 2009; Athanassiadou, 2007). This imposition has been a driving force for considerable research effort, not only in the engineering of UF resins, but also in the development of all sort of alternative resins. In 2007, Dynea AS Company started commercializing AsWood™ resin (formaldehyde based resin), which presented formaldehyde emissions in WBPs similar to the level found in solid wood. However, the price of this product is too high for production of standard particleboard and MDF (Durkic, 2009).

Until now, the decrease on free formaldehyde emissions has been obtained by decreasing the molar ratio F/U and/or by the addition of formaldehyde scavengers. Both lead to a decrease on reactivity and degree of curing, harming the formation of adhesive bonds.

Moreover, currently used hardeners are adapted to high F/U molar ratios and high levels of free formaldehyde in solution. Therefore, the decrease in F/U molar ratio can result in panels with low mechanical performance. The experience of WBP producers is that resins with lower molar ratio F/U are less adaptable to different panel production process conditions and raw materials. This is an important factor, since WBP production nowadays uses mixtures incorporating recycled wood and wood from different origins.

2.2. Melamine-formaldehyde Resins

Melamine-formaldehyde (MF) resins are used mainly as paper impregnating polymers for surfacing of wood-based panels (particleboard and MDF) and decorative laminate. These resins are also used as adhesives to produce particleboard, MDF and plywood when moisture resistance is a desired property. The reduced number of applications, shown in Table 3.2, has to do its high cost (Dunky and Pizzi, 2002 ; Dunky, 2003).

Melamine-formaldehyde resins are also used in specially formulated (i.e. alkylated, methylated, butylated, or isobutylated) resin systems to produce highly durable surface coatings. The coating can be either water based or solvent based. During the coating process these resins form efficient cross-linking systems as they react with polyester, acrylics and epoxies. The benefits of melamine cross-linked coatings include better color retention, wear resistance and scratch resistance. The automobile market accounts for about 40% of MF resin consumption in the surface coating market. (Global Insight Inc., 2007).

Table 3.2. Main applications of melamine-formaldehyde resins

Application	%
Wood Composites: adhesives for moisture resistant composite panels (PB and MDF) and for structural panels (e.g. plywood).	3
Decorative and protective laminates: high-pressure decorative laminates and electrical and mechanical grade industrial laminates: countertops, cabinets, furniture, flooring, wall covering, sheathing, automobile interiors	65
Surface coatings: crosslinkers in stoved paint systems in combination with other polymeric materials including alkyds, acrylics, epoxies and saturated polyesters (automobile, metal containers and furniture, coil coating).	31
Others: Textile treatment	1

2.3. Phenol-formaldehyde Resins

In 1909, Leo Bakeland invented the first synthetic thermosetting resin, a phenol-formaldehyde (PF) resin sold commercially as Bakelite. Even though they found very diverse applications in the past, current use is more restricted, mainly due to high cost (Gardziella et al., 2000; Detlefsen, 2002). Table 3.3 lists the main current applications (Gardziella et al., 2000; Dunky and Pizzi, 2002; Detlefsen, 2002; Dunky, 2003). The high thermal stability and fire resistant properties of these resins allows a wide spectrum of uses in automotive and construction industries (Gardziella et al., 2000). The main use is in the manufacture of fiberglass and rock wool insulation. They are also used as adhesives in structural wood-based panels and as binders in fiberglass and mineral wool insulation.

Another major end use of PF resins is high pressure lamination (HPL), either for decorative or industrial laminates. HPL are composed by a sheet of MF-impregnated decorative paper and several sheets of PF-impregnated kraft paper. The laminated paper is then adhered to a substrate material, usually particleboard or plywood, and is used for countertops, furniture tops, cabinet and drawer faces, wall cladding, automobile interiors, laminated flooring, and wall coverings.

Phenol-formaldehyde resins are also widely used to produce softwood plywood for severe service conditions, oriented strand board (OSB), and particleboard and MDF with high moisture resistance and low formaldehyde emission (Gardziella et al. 2000; Dunky and Pizzi, 2002; Detlefsen, 2002; Dunky, 2003).

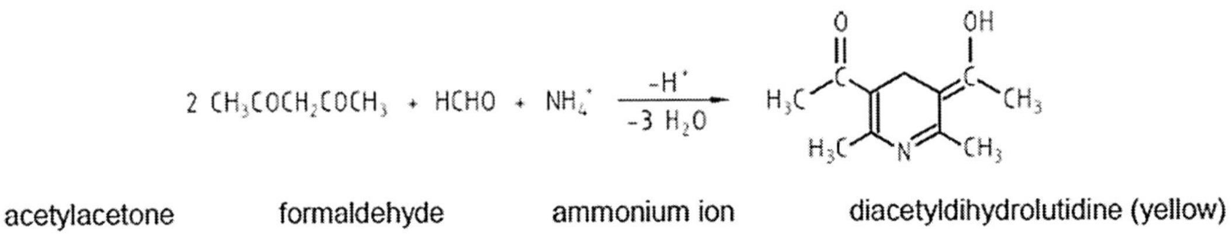

Figure 4.1. Reaction of acetylacetone with formaldehyde (adapted from EN 717-1).

Table 3.3. Main applications of phenol-formaldehyde resins

Application	%
Wood Composites: adhesives for composite panels (PB and MDF) and for structural panels (e.g plywood, OSB, LVL).	20
Decorative and protective Laminates: Resins for high-pressure decorative laminates (countertops, cabinets, furniture, flooring, wall covering, sheathing, automobile interiors) and electrical and mechanical grade industrial laminates.	10
Fiberglass and mineral wool insulation: bind fibreglass, mineral wool or shredded waste products for structural and acoustical insulation, specific applications include low-density insulation, high-density industrial insulation, and other specialty insulation.	30
Abrasive materials: bonded and coated abrasives.	6
Foundry materials: sand binder to coat sand which is then used in core making for casting operations in the foundry industry.	5
Others: Molded plastic products, saturating applications, protective surface coatings, fiber reinforced plastic applications, foam insulation, etc	29

3. Formaldehyde Emissions

3.1. Causes of Emissions

Oxidation of biogenic and anthropogenic hydrocarbons is a source of outdoors formaldehyde emissions. However, exposure to formaldehyde is higher indoors than outdoors due to low air exchange rates (Salthammer et al., 2010). Possible sources of formaldehyde in indoor environments are wood-based materials, insulation materials, coatings, textiles, flooring materials, etc.

Formaldehyde is one of the main components in aminoplastic and phenoplastic resins used in the manufacture of wood-based panels. In board production, formaldehyde can be emitted from the wood raw materials during drying. In the subsequent hot-pressing process, formaldehyde is released from the glue resin and evaporated together with steam (Dunky, 2004). After panel manufacture, formaldehyde emissions during service life are originated not only in residual gas trapped in the substrate structure, but also in formaldehyde dissolved in water present within the board (moisture), and in the hydrolysis of weakly bound formaldehyde from N-methylol groups, acetals and hemiacetals and methylene ether bridges (Dunky, 1998). After panel manufacture, formaldehyde emissions during service life are originated not only in residual gas trapped in the substrate structure, but also in formaldehyde dissolved in water present within the board (moisture), and in the hydrolysis of weakly bound formaldehyde from N-methylol groups, acetals and hemiacetals and methylene ether bridges (Dunky, 1998).

Formaldehyde release from finished panels depends on internal and external factors. The firstfirst include the type of wood and resin used, parameters and operating conditions during panel production, and panel age. External factors are temperature, humidity, air exchange rate, and the total exposed panel area in relation to the total volume of the space in which the panels are placed (Athanassiadou and Ohlmeyer 2009). Test methods for the determination of formaldehyde emission should take into account the factors listed above, in order to be reliable and reproducible.

3.2. Formaldehyde Analysis

Salthammer et al. (2010) present an overview of sampling methods and analytical techniques for the determination of formaldehyde in air. Three main types of methods can be identified: in-situ analysis, derivatization methods and sensor-based methods. For in-situ analysis in outdoor environments, the determination of the concentration of formaldehyde in air is usually made using spectroscopic techniques.. The most popular are Fourier Transform Infrared Spectroscopy (FTIR), but other monitoring techniques can be used as differential optical absorption spectroscopy (DOAS), laser induced fluorescence spectroscopy (LIFS) and tuneable diode laser spectroscopy (TDLS). It is important to take into account the detection limits for these methods (Finlaynon-Pitts and Pitts, 2000). Some of them require long optical paths, which makes the procedure unsuitable for routine applications. Photoacoustics spectroscopy (PAS) can also be used in indoor air.

For indoor applications, derivatization methods are more convenient. Sampling is carried out in batch mode, and formaldehyde from an air stream is trapped in an absorber (generally water, taking advantage of the compound's high solubility) or adsorbed in filters or cartridges. The derivatization reaction results in a chromophore that can be analysed by chromatography and/or spectroscopy. Some of these photometric methods are not formaldehyde-specific, and by-products may interfere in the analysis. The most popular derivatization procedures for formaldehyde analysis are: the chromotropic acid method, the acetylacetone method and the DNPH method, which are briefly summarized below.

In the chromotropic method, formaldehyde reacts with concentrated sulphuric acid (catalyst) and chromotropic acid (1,8-dihydroxynaphthalene-3,6-disulfonic acid) resulting in a red-violet hydroxydiphenylmethane derivative. Then, it reacts with the atmospheric oxygen and gives a violet quinoid. This compound has a maximum absorption at 580 nm and can be detected using a UV/VIS spectrometer (Altshuller et al., 1961). One of the main disadvantages is the low stability of chromotropic acid in solution (Salthammer et al., 2010). An adaptation of the National Institute for Occupational Safety and Health (NIOSH) 3500 chromotropic acid test procedure is used in the large chamber ASTM 1333, small chamber ASTM D 6007 and desiccator ASTM D 5582.

The acetylacetone method (Nash, 1953) is the method recommended by European and Japanese Standards for the determination of formaldehyde content and emission from wood-based panels. Determination is based on the Hantzsch reaction in which formaldehyde reacts with ammonium ions and acetylacetone to yield diacetyldihydrolutidine (DDL) (see Figure 4.1). The determination can be performed by quantitative UV/Vis spectroscopy at 412 nm (DDL has an absorption maximum at 412 nm). The reaction is specific to formaldehyde (EN 717-1).

In European standards EN 717-1, EN 717-2 and EN 120, the formaldehyde solution is mixed with ammonium acetate and acetylacetone solutions and let to react in stoppered flasks during 15 min in a water bath at (40 ± 1) °C. In the desiccator method JIS 1460, the quantities of the reactants are not the same, and the reaction is carried out at (65 ± 2) °C during 10 min. The calibration curve is established from a standard formaldehyde solution. The concentration of formaldehyde is determined by iodometric titration. As DDL also exhibits fluorescense, it can be determined using a fluorimetric spectrophotometer at a wavelength of excitation λ_{ex} = 410 nm and a wavelength of emission λ_{em} = 510 nm. An alternative to acetylacetone has been introduced for derivatization, using acetoacetinalide, which reacts with formaldehyde at room temperature (Li et al, 2007).

The DNPH (2,4-dinitrophenylhydrazine) method is used for the simultaneous analysis of formaldehyde, other aldehydes and ketones (Andrade et al., 1992). In this method, DNPH reacts in acidic solution to give hydrazones, by nucleophilic addition, with liberation of water. The air stream passes through cartridges containing silica gel coated with an acid solution of DNPH. After sampling, the cartridges are eluted with acetonitrile and analysed by HPLC. The separated hydrazones are detected with UV detector (max absorption ranging from 340-427 nm (US EPA Method). It is also accepted by ISO 16000-3. This method can also be used to determine free formaldehyde in phenolic resins (Oliva-Teles, 2002).

Table 4.1. Standards and test methods for the determination of formaldehyde from wood-based panels (Athanassiadou, 2000, Marutzy, 2008)

Test method	Standard, standard draft or method name
Chamber	ASTM E 1333, ASTM D 6007, EN 717-1, JIS A 1901, JIS A 1911, ISO 12460-1, ISO 12460-2
Gas analysis	EN 717-2, ISO 12460-3
Flask method	EN 717-3, método AWPA
Desiccator	ASTM D 5582, ISO 12460-4, JIS A 1460, JAS MAFF 235, JAS 233, AS/NZS 4266.16
Perforator	EN 120, ISO 12460-5
Other	Field and Laboratory Emission Cell "FLEC", Dynamic Microchamber "DMC"

It is important to study the sensitivity (analysis threshold) and specificity of formaldehyde detection for the various methods. Hak et al. (2005) presented an interesting state of the art about the comparison of these methods. They presented an intercomparison of measurement techniques currently used for the detection of atmospheric formaldehyde, as Differential Optical Absorption Spectroscopy (DOAS), Fourier Transform Infra Red (FTIR) interferometry, the fluorimetric Hantzsch reaction technique (five instruments) and a chromatographic technique employing C18-DNPH-cartridges (2,4-dinitrophenylhydrazine).

Other methods for the monitoring of formaldehyde in air are based on sensors. Different kinds of systems have been developed, namely biosensors. However, available sensors have a high detection limit, which makes the technique more suitable for workplace environments (Salthammer et al., 2010). An example of an on-line monitoring system available commercially is the AL4021 by Aerolaser.

3.3. Standard Methods for Emission Testing

The existing methods can be divided in two main categories: measurable emission methods, which determine the actual amount of formaldehyde emitted under the test conditions, and emittable potential methods, which determine the amount of free formaldehyde present in the panel, without considering whether that quantity may actually be released or not, or in how much time (Dunky, 2004). Table 4.1 summarizes the most important test methods and related standards for the determination of formaldehyde from wood-based panels. The methods are described below.

3.3.1. Chamber Method

The evaluation of the real emission of formaldehyde from a product under typical indoor conditions in real-life, and over defined time scales requires the use of a climate-controlled chamber. The formaldehyde concentration in the air inside the chamber is measured along time. The American standard ASTM E 1333 presents a large test chamber that aims to imitate the conditions of a living room with 22 m^2. This test method determines the average formaldehyde concentration in air and emission rate from a number of large size samples under controlled conditions: temperature of (25 ± 1) °C, (50 ± 4) % of relative humidity and air exchange of 0.5/hr. The large test chamber methods, due to their perceived accuracy can be regarded as "standard meter" in formaldehyde testing (Salthammer et al., 2010). However, they are very expensive and time consuming (7 days of conditioning prior testing. The analysis is made at the end of at least a 16 to 20 h period, which could be extended until the formaldehyde concentrations from simultaneous air samples taken from at least two lines do not vary by more than 0.03 ppm. The standard ASTM D 6007 presents a smaller chamber (0.02 a 1 m^3) where the specimens remain until a steady state formaldehyde concentration is reached. The time may be estimated using an equation, which gives a time of analysis around 2.5 hours under the same conditions of ASTM E 1333. Test results in several laboratories indicate a precision of within 0.03 ppm on the same samples in case of ASTM E 1333 and ranging from 0.01 to 0.02 in case of ASTM D 6007. The Californian Air Resources Board (CARB) approved recently regulations that require the use of these chambers for the qualifying tests, which increased the importance of these methods. The International Organization for Standardization (ISO) presents as reference method the standard ISO/FDIS 12460-1 (1 m^3) and a derived method (ISO/DIS 12460-2). The European standard EN 717-1 (chamber method) presents three volume options: > 12 m^3, 1 m^3 and 225 L. The operating conditions are slightly different from the American standard: temperature of (23 ± 0,5) °C and relative humidity of (45 ± 3) %. The air exchange rate is the double of the American standard, i.e 1/hr. The analysis time is at least ten days and the result expressed in $mg.m^{-3}$.

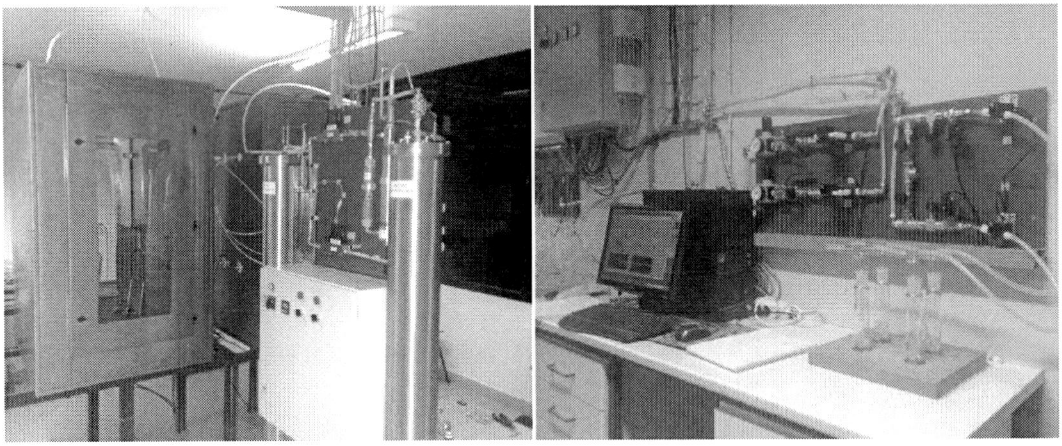

Figure 4.2. Images of small chamber method implementation. Left: 1 m3 chamber according to EN 717-1 and air cleaning and conditioning system. Right: gas sampling system.

The main advantages of the chamber method are the more accurate simulation of the indoor environment and the use of a large volume of sample, which minimizes the influence of material variability. Small chambers, in particular, are currently widely used in Europe and North America and can be very accurate, relatively easy to adapt at both laboratory and plant and correlate well with the large chambers. The formaldehyde concentration is determined by drawing air from the outlet of the chamber through gas washing bottles containing water, which absorbs formaldehyde (Figure 4.2). The concentration of formaldehyde in the chamber atmosphere is calculated from the concentration in water (determined photometrically using the acetylacetone method) and the volume of sampled air. Each of the standards specifies a different procedure for determining when a steady-state condition is achieved. All, however, accept a change in formaldehyde emission of less than 5% over a given period as representing a quasi steady-state condition. In addition, all the standards propose that the test is stopped after 28 days, even if the steady-state condition is not reached (Irle, 2011).

3.3.2. Gas Analysis Method

The gas analysis (EN 717-2) is a derived test that determines formaldehyde release under accelerated conditions: a temperature of 60 °C and within a period of 4 hours. In this method, a test piece with dimensions of 400 mm x 50 mm x board thickness and edges sealed is placed in a closed chamber at (60 ± 0.5) °C with a relative humidity lower than 3 %, an airflow of (60 ± 3) L/h and under an overpressure of 1000 to 1200 Pa. Formaldehyde released from test piece is continually drawn from the chamber and passes though gas wash bottles containing water (Figure 4.3). The formaldehyde is determined at hourly intervals, up to 4 hours. Every hour, the air is automatically led into one of a series of pairs of wash bottles. At the end of the test, formaldehyde release is calculated from the formaldehyde concentration, the sampling time and exposed area of the test piece expressed in mg/m^2h. Even though the time of analysis is short, this test involves a high investment in equipment. The standard EN 13986 indicates this method for faced, coated, overlaid or veneered wood-based panels.

In this method, as well as for the other European methods, the concentration of formaldehyde is determined photometrically (UV/Vis spectrometer) using the acetylacetone method, described above.

Figure 4.3. Gas analysis methodmethod implementation.

Figure 4.4. Implementation of desiccator method.

3.3.3. Desiccator Method

The more relevant desiccator method is defined in the Japanese standard JIS A 1460. It is one of the most economical methods, but it has a drawback. The test pieces shall be conditioned under standard conditions at temperature of (20 ± 2) °C and a relative humidity of (65 ± 5) % until they have attained constant mass, which can take up to one week. Test-pieces are cut into rectangles of 150 mm by 50 mm. A number of test-pieces, corresponding as close as possible to 1800 cm^2 total surface area (ends, sides and faces), are attached to a supporting metal frame and placed on a stainless steel wire net above a crystallizing dish containing water, inside a desiccator with a nominal dimension of 240 mm (Figure 4.4). The lid is placed on the desiccator and the samples are maintained inside for 24 hours at (20 ± 1) °C. The emitted formaldehyde is absorbed by the water in the crystallizing dish. The concentration of dissolved formaldehyde is then determined photometrically using the acetylacetone method, but the reaction conditions and reagent quantities are different from European standards EN 717-1 and EN 120. The emission of formaldehyde is expressed in $mg.L^{-1}$.

There are several variations of the desiccator method as defined in ASTM D 5582, with some differences: the desiccator diameter (250 mm), and the procedure duration, which is 2 hours. Other standards that are based on the same principle are JAS 233 and JAS 235. A recent harmonized standard was adopted by the International Standardization Organization, as ISO/CD 12460-4.

3.3.4. Flask Method

The flask method was developed in the Fraunhofer Institute for Wood Research WKI by Roffael in 1975. A slight modified version of this method was published as EN 717-3. It is a quick method that is suitable for internal quality control in production lines of wood-based panels. This is a static method that consists in suspending test pieces with a total mass of 20 g in a closed container (flask), containing water (50 mL) and maintained at a (40 ± 1)°C during 3 hours. The formaldehyde content in water is determined photometrically by the acetylacetone method and expressed in (mg/kg dry board). The AWPA (American Wood Protection Association) presents a similar method, with the same principle but with different dimensions of the flask. This method does not have great acceptance by the market, nor is significantly used at industrial or academic level.

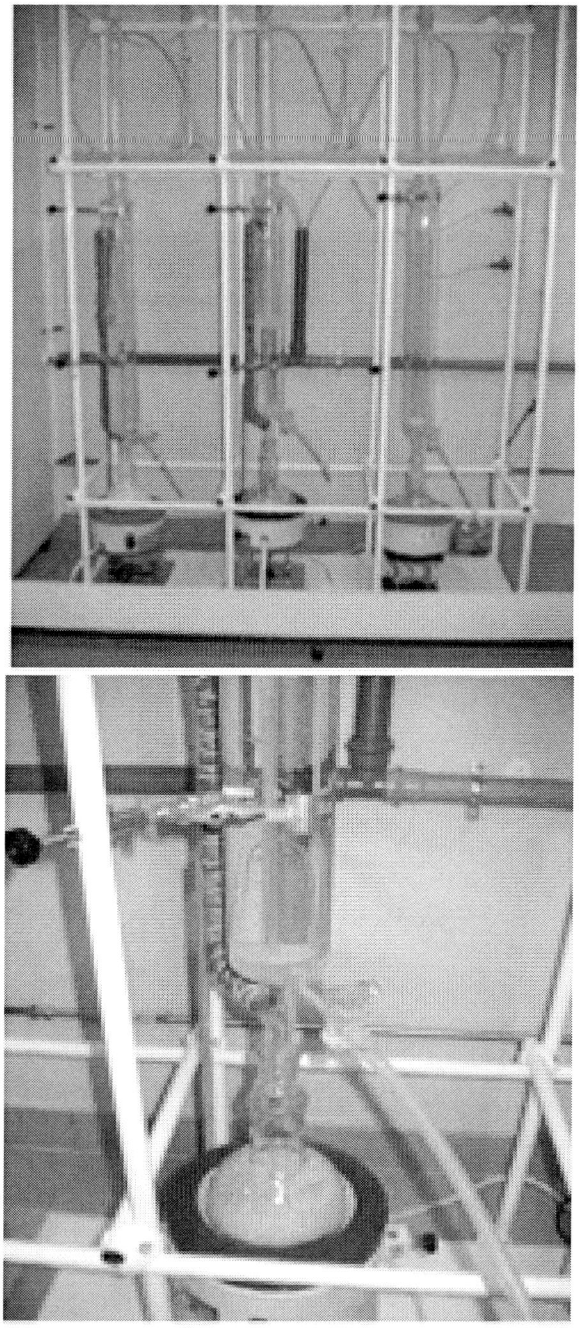

Figure 4.5. Implementation of perforator method.

3.3.5. Perforator Method (Potential Emission)

The perforator method (EN 120) measures the formaldehyde content of wood-based panels and not the actual emission level. While a chamber method test may take several days until the samples attain the equilibrium stage, the perforator method is quicker and expeditious, being indicated for daily factory production control. This is the most popular procedure for measuring formaldehyde content in particleboard and MDF in Europe. EN 13986 indicates this method for unfaced particleboard, OSB, MDF and flaxboards. It is also employed worldwide, except in North America. Formaldehyde is extracted from test pieces (110 g of 25 × 25 mm specimens) by means of boiling in toluene (600 ml), in a round bottom flask connected to a perforator apparatus containing 1000 ml of distilled water. The extraction is carried out during 2 hours, starting at the moment that the first bubbles pass through the filter insert. The water contained in the perforator is, after cooling to room temperature, transferred into a volumetric flask (Figure 4.5). The formaldehyde content of this aqueous solution is determined photometrically by the acetylacetone method. The disadvantage of this method is the environmental impact of the toluene emission and residues. The results are expressed in (mg/100 g oven dry board). The perforator values for particleboards, OSB and MDF shall be applied to wood-based panels conditioned to a reference moisture content (6.5 %.). For different moisture contents, correction factors, calculated by an equation stated in the specifications standards for each type of wood-based panel, are used. This correction factor is contestable as it depends on other factors rather than the moisture content of boards (Roffael and Johnsson, 2011). The accuracy of this method has been very discussed for values below 4 mg/100 g oven dry board. A similar method was established by ISO 12460-5.

3.3.6. Other Methods

There are other methods, used mostly in universities, research laboratories or testing laboratories. One example is the DMC (Dynamic Micro Chamber) used in the United States in factory control quality, but it has not yet been accepted as standard. This method utilizes a combination of a small chamber and electrochemical sensor. It has the advantage of being a short duration test. Another example is the FLEC (Field and Laboratory Emission Cell) implemented for the first time in Scandinavia. In this device, a controlled purified air flow enters the cell and passes through the testing material. The outlet air passes through adsorption tubes, which are connected to a thermal desorption system and analysed in GC/MS or GC/FID system. The great advantage lies in being a transportable emission cell for mobile application (Salthammer et al., 2010). However, a standardized method has not yet been established.

3.3.7. Formaldehyde Methods Survey

Considering the different existing methods, and taking into account that it is difficult to find worldwide agreement on establishing a reference method, it is important to understand the main features of the main methods in use. Table 4.2 summarizes the testing parameters for each method. Table 4.3 surveys the pros and cons for each one.

In reality, no method clearly stands out, all presenting advantages and drawbacks. Implementation costs have been estimated to rate at 0.5:8:100 for perforator, gas analysis and large chamber, respectively (Athanassiadou and Ohlmeyer, 2009). Formaldehyde testing by chamber methods is usually the most time consuming and uses the most sophisticated equipment.

Table 4.2. Testing parameters for the main formaldehyde emission methods

	Methods			
	Chamber EN 717-1	**Gas analysis EN 717-2**	**Desiccator JIS 1460**	**Perforator EN 120**
Pre-conditioning	no	no	7 days	no
Volume	1 m^3	4 L	6 L	-
Temperature	(23 ± 0.5) °C	(60 ± 0.5) °C	(20 ± 0.5) °C	-
Relative humidity	(45 ± 3) %	< 3 %	-	-
Air exchange rate	1 h^{-1}	-	-	-
Loading ratio	1 m^2/m^3	10 m^2/m^3	≈ 30 m^2/m^3	-
Total surface area	1 m^2 (2 boards)	0.040 m^2 (2 boards)	≈ 0.18 m^2	110 g
Unsealed Edges	1.5 m/m^2	no	yes	yes
Testing time	10 to 28 days	4 hours	24 hours	3 hours
Analysis method	acetylacetone	acetylacetone	acetylacetone	acetylacetone
Units	$mg.m^{-3}$ (air)	$mg.m^{-2}h^{-1}$	$mg.L^{-1}$	mg/100 g oven dry board

Table 4.3. Survey of formaldehyde test methods

	Methods			
	Chamber	**Gas analysis**	**Desiccator**	**Perforator**
Pros	- Testing conditions similar to real life - Uses large sample dimensions, which reduces the influence of sample variability	- Short analysis time - Easy to implement	- Low cost equipment	- Short analysis time - Low cost equipment
Cons	- Long analysis time - High cost equipment	- High cost equipment	- Pre-conditioning of samples takes approximately one week	- Toxic waste (toluene) - Low accuracy for very low formaldehyde content

Due to the need of wood-based panel producers to operate in the global market, they have to certificate their products according to the different country or region regulations, which consider different reference methods, like Japan (desiccator), U.S.A. (chamber method) and Europe (perforator method). A new approach for a closer co-operation between different world regions with regard to formaldehyde release test methods was taken between CEN/TC 112 and ISO/TC 89. A resolution was taken in Sydney, 2011-2: "ISO/TC 89 unanimously supports a further development of the standards series ISO 12460 "Determination of formaldehyde release" under the Vienna Agreement in cooperation with CEN/TC 112 to become EN ISO standards".

3.4. International Regulations and Requirements

3.4.1. Formaldehyde Classification

Adverse health effects from exposure to formaldehyde in pre-fabricated houses, especially irritation involving eyes and upper airways, were the first reported in the mid-1960's (Salthammer et al., 2010). In 2004, the International Agency for Research on Cancer (IARC), from World Health Organization, recommended the reclassification of formaldehyde as "carcinogenic to humans (group 1)". In 2006 this recommendation was finally published (IARC, 2006). As a consequence, various authorities and institutions have been concerned about formaldehyde as an indoor pollutant and new regulations have emerged involving increasingly lower exposure limits. Within the European Union, formaldehyde is currently classified as a 3-R40 substance ("limited evidence of carcinogenic effect"), but the classification is being reviewed under the new regulation for chemicals "Registration, evaluation, authorization and restrictions of Chemicals" (REACH). For this purpose, FormaCare (formaldehyde sector group of the European Chemical Industry Council) established a REACH taskforce to facilitate the creation of a consortium allowing European formaldehyde manufacturers to work together as a unified group for their REACH compliance activities.

3.4.2. Occupational Exposure Limits

In the last decades, governments and industry have made considerable efforts to reduce exposure to formaldehyde. The limit levels are separated into two groups: workplace environments (i.e. occupational), and non-occupational environments (i.e. residential) (Salthamer et al., 2010). The occupational exposure limits (OELs) for formaldehyde are separated into three categories: time-weighted average (TWA), short-term exposure limit (STEL) and ceiling limit (exposure limit which should not be exceeded at any time). These limits are different for each country, as seen in table 4.4. For working place thresholds, there are various limits in different European countries, as can be observed in this table. For living room thresholds, recommendation of the German Federal Health Agency in 1977 was 0.1 ppm. Countries with higher limits were compelled to follow the recommendations of IARC (Dunky, 2001). In 1987, the US Occupational Safety and Health Administration (OSHA) established a federal standard that reduced the amount of formaldehyde to which workers can be exposed over an 8-hour work day from 3 ppm to 1 ppm. In 1992, the formaldehyde exposure limit was further reduced to 0.75 ppm. For indoor domestic exposure, the World Health Organisation (WHO) still recommends a limit for formaldehyde air concentration of 0.1 mg/m^3 (for short and long-term exposure) from all sources combined (at this level or below, transient sensory effects should be avoided) (EPF, 2010). The use of the short-term

(30 minute) guideline of 0.1 mg/m^3 (0.08 ppm) also prevents long-term health effects, including cancer (WHO, 2010). The short term exposure levels are associated with acute health effects on individuals, while long-term exposure is related to chronic health effects (Salthammer et al., 2010; Blair 1986).

An EC funded project involving EPF (European Panel Federation) and CEIBois launched a European formaldehyde-in-air monitoring campaign within the wood-based panel manufacturing industry (EPF, 2010). Five small to medium sized manufacturing companies of considerably different ages, located in France, Germany, Poland, Spain and the UK, were selected. Site work was conducted over the 3-week period Wednesday 30th September to Saturday 17th October 2009. In this study, TWA exposure values for a press operator ranged from 0.017 to 0.176 mg/m^3, for a press cleaner TWA ranged from 0.311 to 0.766 and STEL from 0.130 to 1.667, and for a press inspector STEL ranged from 0.183 to 1.187.

3.4.3. Emission Limits for Wood-based Products

In recent years, national regulations for formaldehyde were established and/or reformulated in some countries, limiting formaldehyde emission levels from wood-based panels. The standards for formaldehyde test methods do not refer to a classification of wood-based panels according to the results of formaldehyde emission or release. This classification is established in the specification standards of each product. Table 4.5 lists the current specifications.

The harmonized European standard EN 13986 ("Wood-based panels for use in construction") classifies formaldehyde emission into two classes: E1 and E2. Internal discussions within the European wood-based panel associations, lead EPF (European Panel Federation) to launch its own formaldehyde standard, EPF-S, that corresponds to a perforator value below 4 mg/100 g oven dry wood for PB and 5 mg/100 g oven dry wood for MDF (thickness > 8 mm). Driven by IKEA (IOSMAT 0003), an equivalent class with half E1 formaldehyde emission limits has been introduced: the so-called E0 (or E0.5) (not yet recognized officially by CEN - European Committee for Standardization). Recently, the members of EPF agreed to only produce E1 class, abandoning production of E2 class panels.

In Japan, more strict limits are defined in standards JIS A 5908 e 5905 as, by descending order of emission level, F**, F*** e F****. The F** is more or less equivalent to European E1 class, while the F*** and F**** are much lower. F**** is close to the emission of solid untreated wood, between 0.5 - 2 mg/100 g (Athanassiadou and Ohlmeyer, 2009).

Limits for formaldehyde emission in the United States are described by ANSI A208.1 & 2. More recently, CARB (California Air Resources Board) established more stringent formaldehyde limits for wood-based panels, being nowadays as reference for the wood-based panels market. Phase 1 limits are roughly equivalent to E1 (and F**) class, while Phase 2 limits are similar to F***. These regulations state that, beyond the compliance of those emission limits, wood-based panels and finishing goods for sale or used in California must also be certified by a CARB approved third party certification laboratory, unless they are approved Ultra Low Emission Formaldehyde (ULEF) or No Added Formaldehyde (NAF) products. NAF and ULEF products must demonstrate a 90% or better compliance with a 0.04 ppm (ASTM E1333) limit.

Recently, a new important challenge has been recently imposed by LEED (Leadership in Energy and Environmental Design®) certification, implying the absence of adhesives with urea-formaldehyde chemical bonds in "Green Building" construction (LEED, 2011). Minimizing indoor air contamination associated to substances that are odorous, irritating and/or harmful to the comfort and well-being of installers and occupants is one of the objectives of this organization.

Table 4.4. Occupational Exposure Limits (OELs) for formaldehyde (IARC, 2006, FormaCare 2007, Q&A on formaldehyde, * Decreto Lei 79/2006), adapted from (Athanassiadou and Ohlmeyer, 2009)

Country	Concentration (ppm)	Type
Australia	1.0	TWA
Austria	0.3	TWA
Belgium	0.3	Ceiling
Brazil	1.6	Ceiling
Canada - Alberta	2.0	Ceiling
Canada- Ontário	0.3	Ceiling
Canada - Quebec	2.0	Ceiling
Denmark	0.3	TWA and STEL
Finland	0.3	TWA
France	0.5	TWA
Germany	0.3	TWA
Greece	2.0	TWA
Hong Kong	0.3	Ceiling
Ireland	2.0	TWA
Italy	0.3	Ceiling
Japan	0.5	TWA
Mexico	2.0	Ceiling
Netherlands	1.0	TWA
New Zealand	0.5	Ceiling
Norway	0.5	TWA
Portugal*	0.08	Ceiling
South Africa	2.0	TWA
Spain	0.3	STEL
Sweden	0.5	TWA
Switzerland	0.3	TWA
United Kingdom	2.0	TWA
USA - ACGIH	0.3	Ceiling
USA - NIOSH	0.016	TWA
USA - OSHA	0.75	TWA

TWA –time weight average, STEL – short term exposure limit.

3.5. Correlation between Different Testing Methods

Different authors have attempted to establish correlations between formaldehyde testing methods (desiccator, perforator and chamber). Due to the different operating conditions used in each method, it is not possible to obtain a direct relation, although approximate correlations can be found in literature (Risholm-Sundman et al. 2006; Que and Furuno 2007; Park et al. 2010). In the very low emission range, correlation between corrected perforator values and the real emission of boards is poor (Roffael and Johnsson, 2011). According to these authors, since the mass transfer coefficient is not considered in the perforator method, boards with the same emission value but with different densities may have different real emission characteristics. Table 4.6 presents the transposition of standard limit values to different test methods.

4. Industrial Approaches for Reduction of Formaldehyde Emissions

4.1. Low Formaldehyde Content Resins

Formaldehyde-based resins are still the preferred type of adhesive for industrial production of wood based panels. The most widely used are urea-formaldehyde (UF) resins, followed by phenol-formaldehyde (PF) and melamine-formaldehyde (MF). The industrial success of UF resins is due to the combination of low cost with high reactivity and good physic-mechanical performance.

During service life, formaldehyde emissions (FE) from panels bonded with UF resins can have two origins, besides wood itself: release of unreacted formaldehyde monomer (adsorbed within wood, dissolved in entrapped moisture, or retained in interparticular void space), and long-term resin degradation due to hydrolysis of weak of covalent bonds (Dunky, 2003). Aminomethylene bonds in UF resins are particularly susceptible to hydrolytic attack under humidity conditions. PF resins, on the other hand, are highly resistant to hydrolysis and present much lower formaldehyde emissions after cure. However, the higher cost and lower reactivity imply that PF resins are used mainly in applications implying exterior weather exposure.

In face of increasingly restrict regulations, the initial approaches to reduce formaldehyde emissions in UF resins focused on decreasing the formaldehyde/urea molar ratio (F/U) in synthesis formulations (Myers, 1989). In the last decades, F/U values in resins for WBP production have decreased from about 1.6 to a range between 0.9 and 1.1. The effects of this strategy are well documented (Myers, 1984; Park et al., 2006; Que et al., 2007). In parallel with significant FE decrease, several WBP properties are penalized: internal bond strength, thickness swelling, and water absorption. This lower performance can be compensated by increasing resin dosage, affecting panel cost. In addition to F/U ratio, the synthesis process has a relevant role in the final resin properties, including formaldehyde emissions. Identification of the most favorable reaction conditions and pathways is therefore essential for optimizing the overall performance of the resin (Ferra et al., 2012; Costa et al., 2012; Ferra et al., 2010).

One strategy to counteract the negative effects of decreasing F/U ratio is resin modification with co-monomers, like melamine or phenol. UF resins fortified with relatively small melamine content are a common approach nowadays (Sun et al., 2010; Paiva et al., 2011). These present good mechanical properties and higher resistance to hydrolysis, and hence lower formaldehyde emissions, due to the stability of the bond between methylene and amide group from melamine ring. Due to the much higher cost of melamine in relation to the other monomers, incorporation levels are kept below 5 %. Other co-monomers can be used to attempt a good balance between mechanical properties and formaldehyde emission, like resorcinol, diisocyanates and succinaldehyde (Basta et al., 2006).

Formaldehyde-free urea resins have been reported (Despres et al., 2010), based on dimethoxyethanal, a non-volatile and non-toxic. However, reactivity is much lower than for conventional formaldehyde-urea resins. Combination with about 20 % isocyanate (pMDI) is necessary in order to reduce pressing time and obtain good panel properties.

Table 4.5. Overview on current upper limits of formaldehyde emission (PB – Particleboard, MDF – Medium Density Fibreboard, PW – Plywood, OSB – Oriented Strand Board, LVL – Laminated Veneer Lumber). Adapted from Athanassiadou et al. (2007)

Region	Standard	Test method	Board class	Board type	Limit value
Europa	EN 13986	EN 717-1	E2	PB, OSB and MDF (unfaced)	> 0.124 mg/m^3 air
		EN 120			8 < mg/100 g oven dry board ≤ 30
		EN 717-1		PW, SWP and LVL (unfaced) PW, PB, OSB, MDF, LVL (and others) overlaid	> 0.124 mg/m^3 air
		EN 717-2			3.5 < mg/m^2.h ≤ 8
		EN 717-1	E1	PB, OSB and MDF (unfaced)	≤ 0.124 mg/m^3 air
		EN 120			≤ 8 mg/100 g oven dry board
		EN 717-1		PW, SWP e LVL (unfaced) PW, PB, OSB, MDF, LVL (and others) overlaid	≤ 0.124 mg/m^3 ar
		EN 717-2			≤ 3.5 mg/m^2.h
Japão	JIS A 5908 & 5905	JIS A 1460	F^{**}		≤ 1.5 mg/L
			F^{***}		≤ 0.5 mg/L
			F^{****}		≤ 0.3 mg/L
USA	ANSI A208.1 & 2	ASTM E1333 (large chamber)		PB, MDF	≤ 0.3 ppm
				PW	≤ 0.2 ppm
	CARB	ASTM E1333	Phase 1	PB	0.18 ppm
				MDF	0.21 ppm
			Phase 2	PB	0.09 ppm
				MDF	0.11 ppm

Concomitantly with decreasing F/U ratio, some key variables related to the WBP production process must be taken into consideration in order to minimize formaldehyde emission during the subsequent panel's usable life. Some key variables are (Dunky et al., 2001):

- Moisture content of wood particles or fibers. Higher moisture content usually implies higher FE, either due to retention of dissolved formaldehyde, less effective cure or higher hydrolysis rate.
- Press temperature and press time. Higher cure temperatures and/or times imply higher reaction extension, therefore residual free formaldehyde is decreased and FE will be lower.
- Resin content (gluing factor). Even though higher emissions may be expected from higher resin content in the panel, if higher panel density is obtained then FE tends to decrease. The more tightly packed structure decreases the rate of emission.

Table 4.6. Relationship between different methods and standard limits ([a]Values obtained by correlation) Adapted from Harmon (2008))

Method	Japan		Europe	IKEA	USA	
	F***	F****	E1	E0.5	CARB F1	CARB F2
EN 120 (mg / 100 g odb)	≤ 4.5[a]	≤ 2.7[a]	≤ 8.0	≤ 4.0	≤ 11.3[a]	≤ 5.6[a]
EN 717-1 (mg / m^3 air)	≤ 0.054[a]	≤ 0.034[a]	≤ 0.124	≤ 0.050	≤ 0.176[a]	≤ 0.088[a]
ASTM E1333 (ppm)	≤ 0.055[a]	≤ 0.035[a]	≤ 0.127[a]	≤ 0.051[a]	≤ 0.180	≤ 0.090
JIS A 1460 (mg / L)	≤ 0.5	≤ 0.3	≤ 0.9[a]	≤ 0.4[a]	≤ 1.3[a]	≤ 0.6[a]

4.2. Formaldehyde Scavenger Additives

Formaldehyde scavengers, capable of capturing formaldehyde either physically or chemically and forming stable products, are added to UF resins or to wood particles before pressing. These additives should provide long-term FE reduction, in principle along the panel's service life. Examples used in industry include addition of urea in aqueous solution or powder form, organic amines, scavenger resins (like UF resins with F/U well below 1.0), sulfites, and functionalized paraffin waxes.

The fact that the scavenger inevitably reacts with formaldehyde during pressing, and not only after panel manufacture, has usually a negative effect on bond strength and other properties, since less formaldehyde will be available for the cure reaction. The performance of the panels produced has therefore to be taken into consideration when a scavenger is used. Interestingly, on the other hand, a recent work (Hematabadi et al., 2012) reported that pre-treatment of wheat straw particles with urea solution at 95 °C yielded panels with better mechanical and physical properties, in addition to FE reduction. This was attributed to reaction of free formaldehyde with penetrated urea, resulting in improved bonding performance.

In addition to the cases mentioned above, other formaldehyde scavengers have been reported in the literature, with varying effectiveness. Porous adsorbers like pozzolan and charcoal have been shown to possess some scavenging capability (Kim et al., 2009; Kim et al., 2006). Proteins present in pulp and paper sludge were associated to FE reduction but WBP performance was penalized (Migneault et al., 2011). Addition of tannin solution of UF resin lead to significant FE decrease, due to the reactivity of the hydroxyl groups towards formaldehyde, but caused some reduction in internal bond strength and increased water absorption (Boran et al., 2011a). Very good FE reduction is obtainable by adding sodium metabissulfite to the resin, with no negative impact on other panel properties, but safe handling of this material implies encapsulation (Sene, 2009). Positive results have been reported with several amine solutions besides urea (propylamine, methylamine, ethylamine, and cyclopentylamine), with improvement in physical-mechanical properties in addition to reduction in FE (Boran et al., 2011b). Good results have also been reported for use of different starch derivatives as scavengers (Basta et al., 2006). A recently published work (Zhang et al., 2011) presented significant improvements in bond strength and FE emission in panels prepared with UF resin containing nano-crystalline cellulose previously amino-functionalized with an alkoxysilane.

4.3. Post-treatments

Post-treatments for FE reduction are applied after pressing. Currently used methods include panel impregnation with formaldehyde scavenging species, like aqueous solutions of ammonia, ammonium salts, or urea (Sene, 2009; Dunky et al., 2001). Use of ammonia, however, tends to be abandoned due to toxicity concerns.

Another strategy is the creation of diffusional barriers in the panel surfaces that keep formaldehyde confined. This approach takes advantage of the fact that WPB finishing usually implies application of a laminate, overlay or coating in order to obtain the final decorative appearance. This includes the use of paints, varnishes, veneers, laminates, or resin-impregnated papers. A few works in the literature compare the effectiveness of different barrier materials on FE reduction (Lee et al., 2011; Barry et al., 2006; Composite Panel Association, 2003; Myers, 1986). Epoxy powder coatings and laminate finishes usually imply the highest reduction levels, above 90 %. Combination of liquid coatings with formaldehyde scavenging additives can significantly improve FE reduction. It must be noted, that emissions of other volatile organics (VOCs), in addition to formaldehyde, must also be considered when using coatings.

4.4. Alternative Adhesives

4.4.1. Polyisocyanates

Isocyanate-based adhesives can be used instead of formaldehyde-based resins in production of WBPs. The most common material is pMDI, a complex mixture of the three isomers of methylene diphenyl diisocyanate (MDI), tri-isocyanates and higher polymeric species. pMDI is used either in solvent free form or as emulsion in water (EMDI) (Papadopoulos et al., 2002). Isocyanates react with hydroxyl groups in lignocellulosic wood fibers and with entrapped water (moisture), creating a strong and water-resistant cross-linked structure. Advantages of pMDI for WBP production include: excellent hydrolysis resistance, no volatile emissions after cross-linking, good substrate wettability and penetration, good reactivity (may be increased by addition of catalysts), and excellent mechanical properties at low adhesive contents. On the other hand, several limitations can be identified: high cost in relation to UF resins, need for efficient gas extraction in industrial use, and demoulding difficulties due to adhesion to metal surfaces (Stöckel et al., 2011; Sene, 2009; Dunky, 2003).

Isocyanate-only adhesives are used industrially for production of particleboards, MDF and OSB, but consumption is still much lower in relation to formaldehyde-based resins, mainly due to economic reasons. Hybrid UF-isocyanate adhesives are also used in industrial WBP production. These are obtained by mixing UF resins with lower amounts of pMDI, yielding a copolymerized structure upon cure, with improvements in physical-mechanical properties and formaldehyde emissions (Wang et al., 2004; Simon et al., 2002). The applicability of this type of approach is determined mainly be economic factors.

4.4.2. Natural Adhesives

Industrial use of adhesives obtainable from natural resources (also called bioadhesives or bioresins) has been researched since de 70's, but industrial implementation is still restricted. Production costs, limited availability and consistency of raw materials, and land use issues have been the limiting factors. Advantages of natural adhesives include lower toxicity, biodegradability and production from renewable resources (Dennis, 2007; Dunky, 2003). Three materials have found some success in industrial applications: tannins, lignins and vegetable proteins.

Tannins are polyphenolic compounds obtainable by extraction from wood, bark, leaves, and fruits. Tannin industrial extraction and use is performed almost solely in the Southern hemisphere, using mostly bark from Mimosa, Quebracho and Radiata Pine (Kim, 2009; Dunky, 2003). Use as adhesives implies addition of a hardener, usually formaldehyde. Low FE tannin adhesives are commercially available, but in face of pressure to reduce use of formaldehyde-based adhesives, non-aldehyde hardeners (like hexamine) and autocondensation processes have been investigated, with apparent success (Dennis, 2007; Pizzi, 2006). Addition of tannins to UF and PF resins was reported to reduce FE without impairing mechanical performance (Moubarik et al., 2010). Combination of tannin-formaldehyde adhesives from different origins with poly(vinyl acetate) (PVAc) resins was found to improve bond strength and reduce FE (Kim, 2009; Kim, 2010).

Lignins are abundant phenolic natural polymers that confer mechanical stability to plants, by crosslinking cellulosic components of cell walls. They are obtainable as byproducts of wood pulping. Unlike tannins, there is not a fixed molecular structure attributable to lignins. Composition varies widely depending on the source. Low reactivity is a major disadvantage of its use an adhesive in pure form (Pizzi, 2006; Dunky, 2003). The most interesting potential application is partial substitution of phenol in PF resins, but does not have relevant industrial impact (Sene, 2009; Dennis, 2007; Dunky, 2003).

Soy protein is obtained from soybean, and has been used for centuries as a wood adhesive. In the context of WBP production, soy protein has been added to PF resins to lower

FE, but lower water resistance is an important limitation (Sene, 2009). Formaldehyde-free WBPs have been obtained using an adhesive based on soy flour and glyoxal – a non-toxic, but less reactive, aldehyde (Amaral-Labat, et al. 2008). Use of soy protein combined with polyamidoamine-epichlorohydrin (PAE) resins yields a strong and water resistant product that is commercially available for wood composites (Sene, 2009; Li et al., 2004). Another interesting formaldehyde–free adhesive system, successfully tested in production of plywood and OSB panels, is based on a combination of soy flour, polyethylenimine, maleic anhydride, and sodium hydroxide (Schwarzkopf et al., 2010).

Table 6.1. Possible future WBP classes concerning formaldehyde emission levels

WBP classes	Formaldehyde emission level	Complying adhesive systems
E0 /Carb II/EPF-S	4-5 mg/100 g oven dry board	UF resin modified with 1 -5 % melamine; molar ratio $F/(NH_2)_2$ between 1.00 to 0.90.
F**** ULEF (Ultra Low Emitting Formaldehyde Resins)	0.3 mg/mL	MUF resin with 5 -10 % melamine; molar ratio $F/(NH_2)_2$ between 0.90 to 0.80.
Natural Wood	0.007-0.0125 ppm	MUF resin with 10 - 16 % melamine; molar ratio $F/(NH_2)_2$ between 0.85 to 0.70.
LEED (Leadership in Energy and Environmental Design)	not specified	MF and PF resins (composite materials must contain no added urea-formaldehyde resins)
NAF (No Added Formaldehyde Resins)	not specified	p-MDI; Soy based Adhesive Technology; Bio-adhesives; acrylic resins

FUTURE PERSPECTIVES

The issue of formaldehyde emissions has just recently stirred the WBP industry, in view of the mandatory VOC emission labeling system imposed by French regulations in 2012. This affects all construction products, flooring and wall surfaces, paints and lacquers used indoors. Formaldehyde emissions are seriously restricted: upgrading the rating from C (lowest) to A+ (highest) implies reducing formaldehyde emission from 120 $\mu g/m^3$ (or greater) to 10 $\mu g/m^3$ (or lower), measured in a ventilated test chamber after 28 days of storage. The measurement procedure is based on ISO 16000 testing method. In the short term, this will imply definition of a new class for formaldehyde emission levels from WBPs within Europe, corresponding to emission levels very similar to the ones already established in Japan and USA, namely classes F**** and Carb II, respectively. One other class must be clearly defined, corresponding to emission levels within the range of natural wood (Schafer and Roffael, 2000). This must take into account that the wood species, and the amount and type of recycled wood, used in panel production can affect "natural" formaldehyde emission significantly (Durkic, 2009; Martins et al., 2007).

Also recently, the California Environmental Protection Agency adopted two new classifications for WBPs produced with two particular kinds of adhesives: no-added formaldehyde resins (NAF), and ultra-low-emitting formaldehyde resins (ULEF). Additionally, the U.S. Green Building Council has defined the Leadership in Energy and Environmental (LEED) rating system for green building construction, which specifies that wood composite materials must contain no added urea-formaldehyde resins.

In this context of more stringent regulations, classes E1 and E2, which are currently still allowed in Europe, China, Australia, and Africa, will be reviewed and probably extinct in 2013-2015. It will also be necessary to clarify the relation between the different methods for emission measurement, in order to uniformize the existing classification systems throughout the world (Japan, Europe, USA, and China, among others).

Table 6.1 presents the WBP classes, concerning formaldehyde emission, that will probably prevail in the near future, as well as the complying adhesives (Roschmann and Käsmayr, 2010; Durkic, 2009, Georgia-Pacific, 2009).

REFERENCES

Altshuller, A., Miller, D., Sleva, S. "Determination of formaldehyde in gas mixtures by the chromotropic acid method", *Anal. Chem*, 33, 621, 1961.

Amaral-Labat, G.A., Pizzi, A., Gonc, A. R., Celzard, A., Rigolet, S., Rocha, G. J. M., Eel, D. E. D. L., Paulo, S. "Environment-Friendly Soy Flour-Based Resins Without Formaldehyde" *J Appl Polym Sci*, 108, 624–632, 2008.

Andrade, J.B., Tanner, R.L. "Determination of formaldehyde by HPLC as the DNPH derivative following high-volume air sampling onto bisulfite-coated cellulose filters", *Atmos. Environ.*, 26A (5), 819-825, 1992.

APA, from *http://www.apawood.org/*.

ASTM D 6007 - 02 *Standard Test Method for Determining Formaldehyde Concentration in Air from Wood Products Using a Small Scale* Chamber.

ASTM D5582 – 00 *Standard Test Method for Determining Formaldehyde Levels from Wood Products Using a Desiccator.*

ASTM E1333 - 10 *Standard Test Method for Determining Formaldehyde Concentrations in Air and Emission Rates from Wood Products Using a Large Chamber.*

Athanassiadou E. T. S.; Markessini, C. Towards composites with formaldehyde emission at natural wood levels. In COST Action E49 *Measurement and Control of VOC Emissions from Wood-Based Panels*, Braunschweig, Germany, 2007.

Athanassiadou, E., Formaldehyde free aminoplastic bonded composites", in Proceedings of the *5th International Conference on Environmental Pollution*, Aristotelian University, pp. 170, Thessaloniki, Greece, 2000.

Athanassiadou, E., Ohlmeyer M. Emissions of Formaldehyde and VOC from Wood-based Panels. COST Action WG3 (E49)-*Performance in use and new products of wood based composites*. M. Fan, M. Ohlmeyer, M. Irle, London, Brunel University Press: 219-240, 2009.

Barry, A. Corneau, D. "Effectiveness of barriers to minimize VOC emissions including formaldehyde" *Forest Prod J*, 56, 38-42, 2006.

Basta, A. H., El-saied, H., Gobran, R. H. Enhancing environmental performance of formaldehyde-based adhesives in lignocellulosic composites, part III□: evaluation of some starch derivatives *Des Monomers Polym*, 9, 325-347, 2006.

Blair, A., P. Stewart, O'Berg, M., Gaffey, W., Walrath, J., Ward, J., Bales, R., Kaplan, S., Cubit, D., "Mortality among Industrial-Workers Exposed to Formaldehyde" *J Natl Cancer I*, 76(6), 1071-1084, 1986.

Boran S, Usta M, Gümüskaya E. "Decreasing formaldehyde emission from medium density fiberboard panels produced by adding different amine compounds to urea formaldehyde resin" *Int J Adhes Adhes*, *31*, 674-678, 2011.

Boran, S., Usta, M., Ondaral, S., Gümüşkaya, E. "The efficiency of tannin as a formaldehyde scavenger chemical in medium density fiberboard" *Composites Part B: Engineering*, DOI:10.1016/j.compositesb.2011.08. 004.

CARB, *A Quick Reference Guide to the California Air Resources Board's Airborne Toxic Control Measure* (ATCM) *for Formaldehyde Emissions* from Composite Wood Products, 2009.

Carvalho, L. M. H. *Estudo da Operação de Prensagem do Aglomerado de Fibras de Média Densidade (MDF)*: *Prensa Descontínua de Pratos Quentes University of Porto*, Ph.mD. Thesis, 1999.

CEI-Bois, *Tackle Climate Change: Use Wood*, 2007.

Chapman, K. M. Improved Resin Performance in MDF Through Development of the Blowline Blending Process in *Proceedings of International Panel Products Symposium,* Llandudno, Wales, UK, 5-6 Outober, 2011.

Composite Panel Association, Technical Bulletin - *VOC Emission Barrier Effects of Laminates, Overlays and Coatings for Particleboard,* Medium Density Fiberboard (MDF) and Hardboard 2003.

Costa, N., Pereira, J., Ferra, J., Cruz, P., Magalhães, F.D., Mendes, A., Carvalho, L. "Alternative to latent catalysts for curing UF resins used in the production of low formaldehyde emission wood-based panels" *International Journal of Adhesion and Adhesives*, 33, 56-60, 2012.

Dennis, J. Review of existing bioresins and their applications *Report commissioned from Building Research Establishment,* 2007.

Despres, A., Pizzi, A., Vu, C., Delmotte, L. "Colourless formaldehyde-free urea resin adhesives for wood panels" *European Journal of Wood and Wood Products*, 68, 13-20, 2010.

Detlefsen, W.D. "Phenolic resins: some chemistry, technology, and history ", in: Adhesive Science and Engineering – 2: Surfaces, *Chemistry and Applications, Chaudhury, M. and Pocius,* A. V. (eds.), Elsevier, Amsterdam, 2002.

Durkic,C., Bueso, J. and Fliedner, E. Dynea AsWood™ - Technologies for Composite Boards with formaldehyde emissions as defined by nature" in: *International Conference on Wood Adhesives*, C. Frihart (Ed), Lake Tahoe, Nevada, USA, 2009.

Dunky, M. Urea-formaldehyde (UF) adhesive resins for wood, *Int J Adhes Adhes*, 18(2), 95-107, 1998.

Dunky, M. and Pizzi A. Wood adhesives, in: *Adhesive Science and Engineering – 2*: Surfaces, Chemistry and Applications, Chaudhury, M. and Pocius, A.V. (eds.), Elsevier, Amsterdam, 2002.

Dunky, M. Adhesives in the Wood Industry in *Handbook of Adhesive Technology,* Pizzi, A., Mittal, K. (eds.), Marcel Dekker, New York, NY, 2003.

Dunky, M. Challenges with formaldehyde based adhesives, In: *Proceedings of the COST COST E34 Conference, Innovations in Wood Adhesives*, M. Properzi, F. Pichelin and M. Lehmann (eds), pp. 91-106, Bern University of Applies Sciences, HSB, Biel, Switzerland, 2004.

Dunky, M., Grunwald, D., Haelvoet, W. Emissions, in *Wood Adhesion and Glued Products*. WG2: Glued Wood Products, State of the art Report of COST Action E13, Carl Johan Johansson, Tony Pizzi, Marc Van Leemput (eds.), 2001.

EN 120, *Wood-based panels* — Determination of formaldehyde content — Extraction method called the perforator method.

EN 717-1, *Wood-based panels* — Determination of formaldehyde release — Part 1: Formaldehyde emission by the chamber method.

EN 717-2, *Wood-based panels* — Determination of formaldehyde release — Part 2: Formaldehyde release by the gas analysis method.

EN 717-3, *Wood-based panels* - Determination of formaldehyde release - Part 3: Formaldehyde release by the flask method.

EN 13986, *Wood-based panels for use in constructio*n - Characteristics, evaluation of conformity and marking.

EPF, CEIBois, EFBWW Reduction of formaldehyde exposure in the woodworking industries, 2010 from http://www.cei-bois. org/files/ *Reduction_of_Formaldehyde*_EN.pdf.

European Panel Federation, *Annual Report 2004-2005,* 2005.

European Panel Federation, http://www.europanels.org, 2011.

FAO, http://faostat.fao.org/DesktopDefault.aspx?PageID=626&lang=en # ancor, 2011.

Ferra, J. M. M., Henriques, A., Mendes, A. M., Costa, M. R. N., Carvalho, L. H., Magalhães, F. D., Comparison of UF Synthesis by Alkaline-Acid and Strongly Acid Processes, *J Appl Polym Sci*, 123, 1764-1772, 2012.

Ferra, J. M. M., Mena, P. C., Martins, J., Mendes, A. M., Costa, M. R. N., Magalhães, F. D., Carvalho, L. H. Optimization of the Synthesis of Urea-Formaldehyde Resins using Response Surface Methodology, *J Adhes Sci Technol*, 24, 1455–1472, 2010.

Finlanson-Pitts, B., Pitts, J., *Chemistry of the Upper and lower Athmosphere*, Academic Press, San Diego, 2000.

FormaCare, *Wood panels & formaldehyde*. from http://www.formaldehyde-europe.org/fileadmin/formaldehyde/PDF/BS.Wood.pdf, 2010.

Gardziella, A., Pilato L. A., Knop A, Phenolic Resins: *Chemistry, Applications, Standardization, Safety, and Ecology* (2nd completely revised edition), Springer Berlin, 2000.

Georgia-Pacific Impact of Green Building on Wood Adhesives in: *International Conference on Wood Adhesives*, C. Frihart (ed.), Lake Tahoe, Nevada, USA, 2009.

Global Insight Inc., *Economic Primer on Formaldehyde*, March 2006.

Global Insight Inc., *Socio-Economic Benefits of Formaldehyde to the European Union (EU 25) and Norway October 2007.*

Hak, C., Pundt, I., Trick, S., Kern, S., Platt, U., Dommen, J., Ordónez, C., Prévôt, A., W. Junkermann, W., Astorga-Lloréns, C., Larsen, B., Mellqvist, J., Strandberg, A., Yu, Y., Galle, B., Kleffmann, J., Lörzer, J., Braathen, G., and Volkamer, R. Intercomparison of four different in-situ techniques for ambient formaldehyde measurements in urban air, *Atmos. Chem. Phys. Discuss*., 5, 2897–2945, 2005.

Harmon, D. M. CHANGE - Its Challenges and Opportunities, In *Technical. Formaldehyde Conference*, Hannover, Germany, 2008.

Hematabadi, H., Behrooz, R., Shakibi, a., Arabi, M. The reduction of indoor air formaldehyde from wood based composites using urea treatment for building materials *Constr Build Mater*, 28, 743-746, 2012.

IARC, *Monographs on the Evaluation of Cancinogenic Risck to Humans*, World Health Organization - International Agency For Research On Cancer, 88, 2006.

Irle, M. Analysing formaldehyde and how it can be done, *Wood-based panels International*, May 2011.

Irle, M., Barbu M. Wood-Based Panel Technology in *Wood-based Panels*: *An Introduction to Specialists*, H. Thoemen, M. Irle, M. Sernek. (eds), pp. 1-94, Brunel University Press, London, England, 2010.

ISO 12460-1, *Wood-based panels* — Determination of formaldehyde release — Part 1: Formaldehyde emission by the 1-cubic-metre chamber method.

ISO 12460-3, *Wood-based panels* — Determination of formaldehyde release — Part 3: Gas analysis method.

ISO 12460-4, *Wood-based panels* — Determination of formaldehyde release — Part 4: Desiccator method.

ISO 12460-5, *Wood-based panels* — Determination of formaldehyde release — Part 5: Extraction method (called the perforator method).

ISO 16000-3, *Indoor air* — Part 3: Determination of formaldehyde and other carbonyl compounds - Active sampling method.

JIS A 1460 *Building boards* Determination of formaldehyde emission -- Desiccator method

JIS A 5905 Fiberboards

JIS A 5908 Particleboards

Kim S., Kim H. J., Kim H. S., Lee H. H. Effect of bio-scavengers on the curing behavior and bonding properties of melamine-formaldehyde resins *Macromol Mater Eng*, 291, 1027-1034, 2006.

Kim S. The reduction of indoor air pollutant from wood-based composite by adding pozzolan for building materials *Constr Build Mater*, 23, 2319-2323, 2009.

Kim, S. Environment-friendly adhesives for surface bonding of wood-based flooring using natural tannin to reduce formaldehyde and TVOC emission *Bioresource technol*, 100, 744-748, 2009.

Kim, S. The reduction of formaldehyde and VOCs emission from wood-based flooring by green adhesive using cashew nut shell liquid (CNSL) *Journal Hazard Mater,* 182, 919-22, 2010.

Krüger, U., Kraenzmer, M., Sytrindehag, O. Field studies of the indoor air quality by photoacoustic spectroscopy, *Environment International*, 21, 791, 1995.

Lee, J.-hun, Kim, J., Kim, S. Formaldehyde emission and burning behaviors from wood-based panels according to various surface finishing methods *Proceedings of 5th*

International Symposium on Sustainable Healthy Buildings, Seoul, Korea 10 February 2011, pp. 327-334.

LEED. *LEED 2009 For Commercial Interiors Rating System*, 2011.

Li, K., Peshkova, S., Geng, X. Investigation of Soy Protein-Kymene ® Adhesive Systems for Wood Composites *J Am Oil Chem Soc*, 81, 487-491, 2004.

Li, Q., Srithanranthikhun, P., Motomizu, S. Development of Novel Reagent for Hantzsch Reaction for the Determination of Formaldehyde by Spectrophotometry and Fluorometry, *Anal Sci*, 23, 413, 2007.

Marra, A. A. *Technology of wood bonding: principles in practice*, Van Nostrand Reinhold, 1992.

Martins, J., Pereira, J., Pinto, B., Coelho, C., Carvalho, L. H. C., Effect of recycled wood on formaldehyde release of particleboard, COST Action *E49 Conference Measurement and Control of VOC Emissions from Wood-Based Panels*, Fraunhofer WKI, Braunschweig, Germany, 28-29 November 2007, published in CD.

Maruztky, R. Global formaldehyde regulations and requirements: Current situation and developments, *6th European Wood-based Panel Symposium*, Hanover, October 10th, 2008.

Migneault S., Koubaa A., Riedl B., Nadji H., Deng J., Zhang T. Potential of pulp and paper sludge as a formaldehyde scavenger agent in MDF resins *Holzforschung,* 65, 403-409, 2011.

Moubarik, A., Pizzi, A., Allal, A., Charrier, F., Khoukh, A., Charrier, B. Cornstarch-mimosa tannin-urea formaldehyde resins as adhesives in the particleboard production *Starch - Stärke*, 62, 131-138, 2010.

Myers G. E. How mole ratio of UF resin affects formaldehyde emission and other properties: a literature critique *Forest Products Journal,* 34, 35–41, 1984.

Myers, G. E. Effects of post-manufacture board treatments on formaldehyde emission:: a literature review (1960-1984) *Forest Products Journal,* 36, 41-51, 1986.

Myers, G. E. Advances in methods to reduce formaldehyde emission, in *Composite Board Products for Furniture and Cabinet-Innovations in Manufacture and Utilization*, M. P. Hamel (ed.), Forest Products Society, Madison, WI, 1989.

Nash, T. The colorimetric estimation of formaldehyde by means of the Hantzsch reaction, *Biochem J.*, 55, 416, 1953.

Oliva-Teles, M. T., Paíga, P., Delerue-Matos, C. M., Alvim-Ferraz, M. C. Determination of free formaldehyde in foundry resins as its 2,4-dinitrophenylhydrazone by liquid chromatography, *Anal Chim Acta*, 467, 97–103, 2007.

Paiva, T.; Henriques, A.; Cruz, P.; Ferra, M.; Carvalho, L. H. Production of Melamine Fortified Urea-Formaldehyde Resins with Low Formaldehyde, *Emission* 124, 2311–2317, 2012.

Papadopoulos, A. N., Hill, C. A. S., Traboulay, E., Hague, J.R.B. Isocyanate Resins for Particleboard: pMDI vs EMDI *Holz Roh Werkst*, 60, 81-83, 2002.

Park, B.-D., Chang Kang, E., Yong Park, J. Effects of formaldehyde to urea mole ratio on thermal curing behavior of urea–formaldehyde resin and properties of particleboard *J Appl Polym Sci,* 101, 1787-1792, 2006.

Pizzi, A. Recent developments in eco-efficient bio-based adhesives for wood bonding: opportunities and issues *J Adhes Sci Technol*, 20, 829-846, 2006.

Que, Z., Furuno, T. Formaldehyde emission from wood products: relationship between the values by the chamber method and those by the desiccator test, *Wood Science and* Technology 41(3), 267-279, 2007.

Que, Z., Furuno, T., Katoh, S., Nishino, Y. Effects of urea–formaldehyde resin mole ratio on the properties of particleboard *Build Environ*, 42, 1257-1263, 2007.

Risholm-Sundman, M., Larsen, A., Vestin, E., Weibull, A. Formaldehyde emission - Comparison of Different standard methods, *Atmosferic Environment*, 41, 3193-3202, 2006.

Roffael, E., C. Behn, B. Dix and G. Bar. Recycling UF-Bonded Fibreboards in International Panel Products Symposium. M. Spear. Nantes, France, 2009.

Roffael, E., Johnsson, B. The perforator value in balance, *Proceedings of the International Panel* Products *Symposium*, Llandudno, Wales, U.K., 2011.

Roschmann, K., Käsmayr D. Novel Concept for Particle Boards with low Formaldehyde Emissions, in *7th European Wood-Based Panel Symposium*, Hanover, Germany, October, 2010.

Salthammer, T., Mentese, S., Marutzky, R. Formaldehyde in the Indoor Environment *Chem Rev*, 110, 2536–2572, 2010.

Schafer, M., Roffael, E. On the formaldehyde release of wood *Holz Roh Werkst*, 58, 259-264, 2000.

Schwarzkopf, M., Huang, J., Li, K. A Formaldehyde-Free Soy-Based Adhesive for Making Oriented Strandboard *J Adhesion,* 86, 352–364, 2010.

Sellers, T. *Plywood and Adhesive Technology*. New York, Marcel Dekker, 1985.

Sene, M.-L. *Research of alternative solutions able to limit the formaldehyde release in panels (manufacturing & use) Rescoll Centre Technologique*, Pessac, France, 2009.

Sigvartsen, T. and M. Dunky New Adhesive Development to Meet the Challenges of Tomorrow in Wood Adhesives. San Diego, California, USA, 2005.

Simon, C.; George, B.; Pizzi, A. Copolymerization in UF/pMDI adhesives networks *J Appl Polym Sci*, 86, 3681-3688, 2002.

SRI, Urea-Formaldehyde (UF) Resins, *SRI Consulting*, 2009.

Stöckel, F., Konnerth, J., Moser, J., Kantner, W., Gindl-Altmutter, W. Micromechanical properties of the interphase in pMDI and UF bond lines *Wood Sci Technol,* DOI: 10.1007/s00226-011-0432-0, 2011.

Suchsland, O., Woodson, G. E., Fiberboard manufacturing practices in the United States, *USDA Forest Service, Agriculture Handbook*, 1987.

Sun, Q.-ning, Hse, C.-yun, Shupe, T. F. Characterization and Performance of Melamine Enhanced Urea Formaldehyde Resin for Bonding Southern Pine Particleboard *J Appl Polym Sci*, 119,3538–3543, 2011.

Sun, X.S., Soy Protein Adhesives in *Bio-Based Polymers and Composites*, Wool, R.P, Sun. X.S. (eds.), Elsevier Academic Press, Burlington, MA, 2005.

Tang, X., Bai, Y., Duong, A., Smith, M. T., Li, L., Zhang, L. Formaldehyde in China: Production, consumption, exposure levels, and health effects, *Environ Int*, 35, 1210-1224, 2009.

U. S. EPA Method T0-11 A *Determination of Formaldehyde in Ambient Air Using Adsorbent Cartridge Followed by High Performance Liquid Chromatography* (HPLC), U.S. EPA, Cincinnati, 1999.

Wang, W. Zhang, X., Lu, R. Low formaldehyde emission particleboard bonded by UF-MDI mixture adhesive *Forest Prod J*, 54, 36-39, 2004.

Waters, G. D. Medium Density Fibreboard Blowline Blending – Theories in and around the Black Box in *Procedings NPA Resin and Blending Seminar*, Irving, Texas, pp 56-61, 1990.

WHO, WHO *guidelines for air quality: selected pollutants*, WHO Regional Office for Europe, Copenhagen, Denmark, 2010.

Wijnendaele, K. Panel Industry Struggles Through Economic Crises towards Bright Future for Carbon-Storing Panels. *International Panel Products Symposium* M. Spear. Nantes, France, 2009.

Youngquist, J. A. *Wood-based Composites and Panel Products*, F. P. Laboratory, 1999.

Zhang, H., Zhang, J., Song, S., Wu, G., Pu, J. Modified nanocrystalline cellulose from two kinds of modifiers used for improving formaldehyde emission and bonding strength of urea-formaldehyde resin adhesive, *BioResources*, 6, 4430-4438, 2011.

In: Formaldehyde
Editors: Chan Bao Cheng and Feng Hu Ln
ISBN 978-1-62257-214-4
©2012 Nova Science Publishers, Inc.

Chapter 3

ELECTRONIC SPECTRA OF FORMALDEHYDE IN AQUEOUS SOLUTION: THE NONEQUILIBRIUM SOLVENT EFFECT WITH MOLECULAR MODELING

Quan Zhu[1] **and Yun-Kui Li***[2]*
[1]College of Chemical Engineering, Sichuan University, Chengdu, China
[2]College of Enology, Northwest A&F University, Yangling, China

ABSTRACT

Two models are presented to estimate the electronic spectra for formaldehyde in condensed phase. Different from others' concerns, the key of our models is the establishment of proper energy expression to describe the Franck–Condon state with the aid of the constrained equilibrium method under the thermodynamics theory. In the first explicit solvent model, high-level quantum mechanics theory is employed to calculate formaldehyde and classical molecular dynamics method is adopted to simulate the individual solvent molecules. Mutual polarization between the two portions is adequately considered. The long–range electrostatic effect and short–range dispersion/repulsion effect in the solute–solvent system are introduced into the solute Hamiltonian as perturbation operators. In the second implicit solvent model, the solute formaldehyde with the point dipole approximation is located at the center of a spherical cavity surrounded by continuous dielectric medium and the cavity radius is determined by the molecular dynamics simulation. Both the two models well predict the solvatochromic shift of the singlet n $\rightarrow \pi^*$ transition for formaldehyde in aqueous solution. Different contributions to the total solvation shift are analyzed and the main component comes from the electrostatic plus polarization interactions. The microscopic solvent structure is quite disturbed by formaldehyde to form three solvation shells. There are mainly three or two dynamic hydrogen bonds formed between formaldehyde and water molecules. Since the hydrogen bonding effect is always deemed to be the dominant contribution to the

* Corresponding author. Tel.: +86 28 85403231, Fax: +86 28 85407797, E-mail address: qzhu@scu.edu.cn (Q. Zhu)
* Corresponding author. E-mail address: leeyking@gmail.com (Y.K. Li)

solvatochromic shift of polar chromophores in aqueous solution, the lowest singlet n → π* transition spectra of the supermolecular clusters of CH_2O–nH_2O (n=1, 2, 3) are studied based on the structures optimized by quantum mechanics methods or extracted from molecular dynamics simulation. The results are consistent from our two models.

1. INTRODUCTION

As the simplest carbonyl compound, formaldehyde is an interesting model and probe for the studies of organic and biological chromophores in both experimental and theoretical researches. Formaldehyde is an important substance in synthetic chemistry, combustion chemistry, environmental chemistry, atmospheric chemistry and interstellar chemistry. In these fields, the spectrum feature is a significant facet. The excitation of formaldehyde to a higher energy state always couples with a great change of the molecular dipole moment [1], i.e. the molecular polarity of an excited state is different from that of the ground state. A representative electronic excitation for formaldehyde is the lowest singlet n → π* transition, which refers to the promotion of an n-electron from the carbonyl oxygen's lone-pair orbital to the empty carbonyl antibonding π* orbital [1]. This promotion leads to the centralization of electron distribution on the carbonyl group and thus reduces the dipole moment and polarity. When formaldehyde is solvated into polar solvents, it will be stabilized by the solvent differently in the ground versus the n → π* excited state. Polar solvents solvate the formaldehyde at the ground state with a stronger polarity more favorably than that at the n → π* excited state, resulting in an enhancement of the energy difference between the two states. This implies the electronic absorption spectrum for the n → π* transition will shift to a shorter wavelength in polar solvents relative to in vacuum, i.e. blue shift or positive solvatochromic shift (see figure 1).

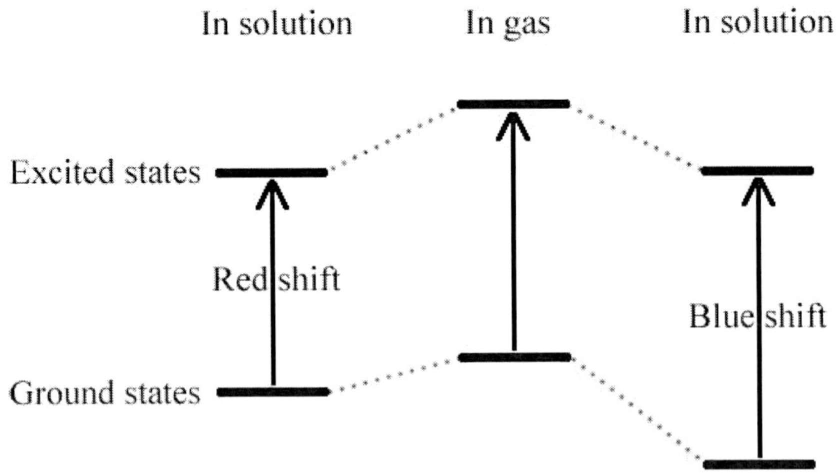

Figure 1. A schematic diagram of the red and blue shifts for the spectra of a chromophore. The length of the arrows indicates the relative magnitude of the vertical excitation energies required for transitions from the ground to excited states.

For theoretical chemists, how to model the electronic spectra in condensed phase has long been a challenging topic. The difficulty mainly comes from the complicated solute–solvent specific interactions, such as hydrogen bonding in hydroxylic solvents, and more generally from the solvent dielectric effects, such as dipole–dipole and dipole–induced dipole interactions, as well as packing and orientation strain in the Franck–Condon excited state [1]. A feasible way to solve this problem may be the reasonable description of the Franck–Condon state in the presence of solvent surrounding. Thus far, numerous models have been proposed to evaluate the solvent effects on spectra in solution.

Generally, the solute molecule is treated by quantum mechanics (QM) theories, while the solvent surrounding is treated by continuous medium theory, Monte Carlo (MC) simulation, molecular dynamics (MD) simulation or statistical-mechanical integral methods. The first one, also called as implicit solvent model (see figure 2), treats the solvent as a uniform continuous medium characterized by the macroscopic static and optical dielectric constants. The treatment focuses on the long–range electrostatic interaction between the solute and solvent and can efficiently reduce the computational cost. However, it lacks the consideration of specific solute–solvent interactions and cannot provide any microscopic detail.

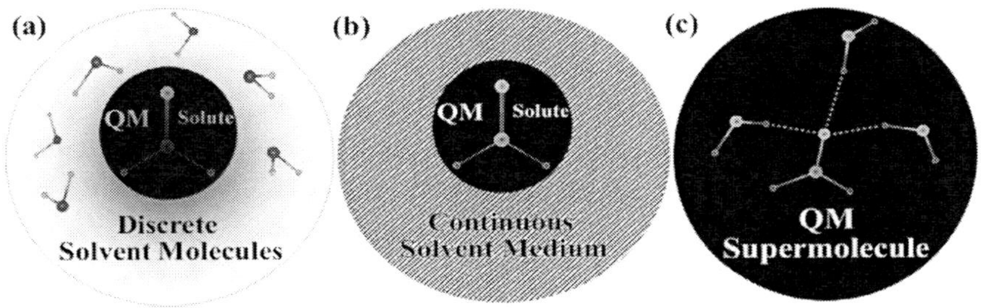

Figure 2. A schematic two-dimension-projection illustration for the modeling of solvent effects on spectra. (a) The explicit solvent models that treats the solute molecule with quantum mechanics theories and the discrete solvent molecules with Monte Carlo, molecular dynamics simulation or statistical-mechanical integral methods. (b) The continuum models that treats the solute molecule with quantum mechanics theories and the solvent with continuous medium theory. (c) The supermolecule model that consider both the solvent molecules in the first solvation shell and the solute molecule with QM theories.

The last three treatments, also called as explicit solvent model (see figure 2), consider the solvent as discrete molecules. They can provide atomic-level structure of the solvent environment, thermodynamic properties, etc. However, for QM/MC or QM/MD method, the computational cost may be very expensive to sample sufficient configurations and perform high-level QM calculations for each configuration. A compromise solution is to average the solvent structure or the potential representing the solvent environment. Another way to solve this problem is to take advantage of the statistical-mechanical theory, which can provide a time saving strategy to get the statistical distribution of solvent molecules at the time of optimizing the electronic structure of solute in solution [2, 3, 4]. The accuracy of these explicit solvent methods is significantly dependent on the potential representing the solvent.

Furthermore, some neighbor solvent molecules can also be clustered with the solute molecule and treated with QM methods, i.e. a supermolecule strategy (see figure 2), since the solvent in the first solvation shell has the greatest influence on the solute if there are strong solute–solvent interactions like hydrogen bonding. In the supermolecule model, the other solvent molecules can be investigated by the continuous medium theory, the MC or MD simulation. As a derivative of the supermolecule strategy, the solute moiety plus partial or entire solvent molecules can be calculated at high-level QM theories while the rest of the solvent molecules is estimated at lower-level QM theories [5], with the structures (configurations) drawn from MD or MC simulation. In Section 2, a brief review is presented about the models proposed to estimate the spectra of formaldehyde in solution. Section 3 gives details of the theoretical derivations of our explicit and implicit solvent models. Calculation details and discussions of the results are given in Sections 4 and 5, respectively. Some conclusions will be drawn in Section 6.

2. Historical Modeling for the Spectra of Formaldehyde in Solution

Specifically for hydrated formaldehyde, a variety of models belonging to the aforementioned three categories have been reported to simulate the solvent effects on the spectra. The following paragraphs provide a brief overview of these models.

2.1. Implicit Solvent Model

When a chromophore solute experiences a Franck–Condon transition, its charge distribution will change rapidly, resulting in a sudden change of the solute field. The responses of the solvent to this field change can be classified into two kinds: the low frequency from the orientation/vibration motions of solvent nuclei, marked as inertial or orientational polarization, and the high frequency from the solvent electronic motions, marked as dynamical or electronic polarization. These two different time-scale responses lead to the nonequilibrium polarization situation related to the Franck–Condon state of the solute. In the equilibrium polarization, the total solvent polarization equilibrates with the solute charge distribution. In the nonequilibrium polarization, the dynamical or electronic polarization can adjust itself quickly to equilibrate with the solute new charge distribution, while the inertial or orientational polarization has no time to respond to the fast change of the solute charge distribution and will keep fixed as the value in the previous equilibrium state.

The primary versions of the energy expression of the nonequilibrium polarization were proposed by Ooshika [6], Marcus [7], Lippert [8], and Mataga et al. [9] in the framework of continuous medium theory in the 1950s. Afterwards, based on the foundation of Felderhof's energy expression for medium polarization [10], Lee and Hynes [11], Kim [12], Aguilar et al. [13] obtained a similar form to that presented by Marcus. However, the traditional expressions seem incapable of explaining some related experimental phenomena [14, 15, 16, 17, 18, 19, 20] and not well-founded according to the basic thermodynamic principles [21].

A few years ago, we threw doubts on the rationality of the original Marcus theory, and argued that the reversible work method employed by Marcus in integrating the free energy along an ultra-fast process which links the ground state equilibrium polarization to the Franck–Condon state nonequilibrium polarization, was inappropriate, since the classical thermodynamics requires a reversible pathway when one tries to gain the free energy change. It is clear that the process starting from the equilibrium polarization to the nonequilibrium polarization is not quasi-static. Hence, the treatment by Marcus should be problematic.

Aguilar and Sanchez et al. [13, 22] proposed a nonequilibrium polarizable continuum model (NE-PCM) to investigate solvent effects on optical emission and absorption spectra of formaldehyde in solution. The response of the medium to the solute is characterized by virtual charges located on the boundary surface of the cavity. Mikkelsen and Jensen et al. [23] proposed the solute–solvent polarizable continuum model of the nonequilibrium multiconfigurational self-consistent reaction field (MCSCRF) method to study the solvatochromic shifts in electronic spectra of formaldehyde in solution. This method uses full multipole expansions for the solute charge. Mennucci et al. [18] utilized the integral equation formalism (IEF) method with a nonequilibrium approach to study excited states and solvatochromic shifts of formaldehyde in water. Andrade do Monte et al. [24] combined the conductor-like screening model (COSMO) with multireference configuration interaction with singles and doubles excitations (MR-CISD) to study the n $\rightarrow$ π^* and π $\rightarrow$ π^* states of formaldehyde in several solvents.

In this model, the solute is confined in a cavity constructed from overlapping atomic-centered spheres of the solute. The effect of the polarized continuum is represented by screening charges on the cavity surface segments. Improta, Barone et al. [25, 26] developed a state specific polarizable continuum model time dependent density functional theory (SS-PCM/TD-DFT) approach to investigate the absorption and emission spectra of formaldehyde in aqueous solution. The state specific approach explicitly considers the variation of the dipole moment associated with electronic transition.

2.2. Explicit Solvent Model

Different from implicit solvent models that emphasize on the low-cost and the description of the solvent from a macroscopic point of view, explicit solvent models pay more attention to the rational representation of the complicated solute–solvent interactions and the microscopic details of the solvent. The development of explicit solvent models mainly focuses on several aspects: (i) a polarizable solvent model to consider solute–solvent mutual polarization; (ii) high-level QM methods to improve the accuracy; (iii) sufficient solute–solvent configurations and the computational expense; (iv) the solute–solvent interaction potential.

Blair et al. [27] employed Hartree-Fock (HF) and MD simulation to study the solvent effect on the $^1A_1 \rightarrow {}^1A_2$ transition of formaldehyde in water. Only the electrostatic interactions between the solute and solvent were taken into account. Fukunaga and Morokuma [28] used the interaction energy method to study the solvent effects on formaldehyde $^1(n, \pi^*)$ transition. They first employed MC simulation to sample the positions and orientations of water molecules around formaldehyde. On the basis of ab initio molecular orbital calculations, they derived potential functions for interaction between formaldehyde and one water molecule.

With the aid of these formaldehyde–water potential functions and the MCY potential function [29] expressing water–water interactions, they calculated the interactions between formaldehyde and water, in both the ground and the nπ^* singlet excited state. Thompson [30] proposed a hybrid quantum mechanical/molecular mechanical model (QM/MMpol) to study the spectroscopy of formaldehyde in water. They employed a polarizable model for the solvent. Afterwards, Kawashima and Dupuis et al. [31, 32, 33] introduced the QM/MM-pol-vib/CAV model to study properties of microsolvated formaldehyde. This model permits polarizable solvent and intra-molecule vibrations of water molecules. The solute-solvent cluster is confined in a spherical cavity and vacuum or dielectric continuum is appended outside.

With employing high-level QM methods, Kongsted et al.[34, 35, 36] presented a combined coupled cluster/molecular mechanics (CC/MM) model; Xu et al. [37] proposed a multireference configuration interaction/molecular dynamics (MRCI/MD) approach; Malaspina et al. [38] developed the sequential classical Monte Carlo simulations and quantum mechanics (S-MC/QM) approach with TDDFT wave function for the solute; and Lupieri et al. [39] presented hybrid ab initio Car Parrinello molecular dynamics/ molecular mechanics (CPMD/MM) with TDDFT and complete active space with second-order perturbation theory (CASPT2) methods for excitation energy calculations to study spectroscopic properties of microsolvated formaldehyde.

In order to improve computation efficiency, Ten-no et al. [2, 3] employed the statistical mechanics integral method of the reference interaction site model (RISM) to obtain the solvent distribution around formaldehyde. This model can provide information equivalent to those obtained from MC or MD simulation and reduces the computational cost drastically by these analytical treatment of statistical mechanics theories. However, this model only take the electrostatic interactions into account. Then, Yoshida and Kato[4] proposed the molecular Ornstein–Zernike self-consistent-field (MOZ-SCF) method, which incorporates the exchange repulsion/charge transfer terms by introducing an effective potential located on solvent molecule. In recent years, Martin et al. [40] exploited the mean field approximation (MFA) to reduce the computational cost in their averaged solvent electrostatic potential/molecular dynamics simulation (ASEP/MD) model.

Öhrn and Karlströmp [41, 42] introduced nonelectrostatic effects of the solvent into the solute Hamiltonian in their explicit solvent QMSTAT model. Gordon et al. [43, 44] proposed the effective fragment potential (EFP) model, which incorporates electrostatic (Coulomb), polarization (induction), exchange repulsion and charge transfer interactions, as well as short–range electron correlation effects to describe intermolecular interactions. This method is combined with the CIS, TDDFT and EOM-CCSD [45] methods to investigate the n $\rightarrow \pi^*$ and $\pi^* \rightarrow$ n fluorescence spectra of formaldehyde in water.

2.3. Supermolecule Model

Several supermolecule models also have been proposed to describe the spectroscopic properties of hydrated formaldehyde. Dimitrova et al. [46] studied the hydrogen–bonded formaldehyde–*n*water (n=1, 2, 3) complexes by means of ab initio SCF and CI calculations. The structures were optimized by QM methods. Canuto and Coutinho [47, 48] performed full quantum mechanical intermediate neglect of differential overlap/singly excited configuration interaction (INDO/CIS) calculations for clusters of formaldehyde and water molecules (15, 35, 80 and 142 waters) to analyze the n $\rightarrow \pi^*$ and $\pi^* \rightarrow$ n transitions. The cluster structures are generated by MC simulation. Hirata et al. [49] proposed a binary-interaction method for clusters of weakly interacting molecules combined with TDDFT and equation-of-motion coupled-cluster singles and doubles (EOM-CCSD) methods. They considered the water molecules as distributed dipoles and studied the CH2O–*n*H2O (n=2, 30, 81) clusters with the geometries optimized in vacuum or extracted from classical MC simulation. Mochizuki et al. [5, 50, 51] applied the multilayer fragment molecular orbital (MLFMO) method to study the first excitation energy for $CH2O$–$n$$H2O$ (n=3, 16, 128) clusters within the QM/QM framework. Formaldehyde plus several water molecules were optimized by QM method and further described with the multiconfiguration self-consistent-field (MCSCF) for excitation calculation, while the other water molecules were optimized by MM method and treated with HF theory for excitation calculation. An extended version of this model is combined with the TDDFT method [52].

2.4. The Aims of this Chapter

Although quite a few models, with emphasis on diverse aspects, have been presented to evaluate the optical spectroscopy properties in solution, the key and foundation is the establishment of a proper energy expression for the Franck–Condon state in the nonequilibrium. Most recently, by introducing the constrained equilibrium approach [53], the novel formula for the electrostatic solvation energy of nonequilibrium polarization has been proposed in the continuum model by our group [21] and applied to electron transfer reactions [54, 55], vertical ionization energy of hydrated electron [56], photoinduced electronic excitation in aqueous solution [57, 58]. In this chapter, the previous formulas for nonequilibrium polarization in continuum model was further equivalently extended to the situation of explicit solvent model with the discrete representation of the solvent dipoles and introduction of the solvent molecular polarizability. The external field for constructing the virtual constrained equilibrium state can be expressed with the physical quantities before and after the Franck–Condon transition. The ASEP/MD program [59] proposed by Aguilar et al., provides a basic frame to apply our new nonequilibrium solvation theory based on the explicit solvent model. This program combines the Gaussian 98 and the Moldy programs to realize the QM/MD strategy. Instead of performing a full polarizable molecular dynamics, it employs the nonpolarizable TIP3P water model to perform MD simulations to sample configurations. To complement the deviation stemming from such nonpolarizable treatment of the solvent, a polarization calculation is carried out sequentially with introducing the solvent molecular polarizability so as to consider the mutual polarization between the solute and solvent. This is economical and has been proved to be equivalent to deal with the solvent polarization, compared to the full polarizable strategy. The usage of the mean field approximation (MFA) reduces the computational cost and makes it possible to describe the solute molecule with high level quantum theories. We modified the nonequilibrium polarization module of ASEP/MD program to implement our new expression in explicit solvent model. The new codes (M-ASEP/MD) are applied to account for the solvatochromic shift of the $n \rightarrow \pi^*$ transition spectra of formaldehyde in water.

With the approximation of point dipole and sphere cavity, the new analytical expression for the solvatochromic shift for formaldehyde is also deduced in continuum model. Formaldehyde is approximated as a point dipole confined in a cavity surrounded by continuous dielectric medium. The solute cavity radius is an important parameter for the evaluation of the shift; however, the traditional determination of the radius lacks consideration of the microscopic structure of the solvent surrounding and specific solute–solvent interactions. Thus, the solute radius is estimated from the solute–solvent configurations extracted from the MD simulations in this work, and the corresponding prediction for the solvent shift is very close to that obtained from the M-ASEP/MD program. We further present a supermolecule model to study the solvent effects on the lowest singlet n → π* transition spectra of formaldehyde in aqueous solution, since the hydrogen bonding effect is always deemed to be the dominant contribution to the solvatochromic shift of polar chromophores in polar solvents. The MD simulation shows the microscopic solvent structure is quite disturbed by formaldehyde to form three solvation shells. There are mainly three or two dynamic hydrogen bonds formed between formaldehyde and water molecules. Thus, the clusters of CH_2O–nH_2O (n=1, 2, 3) are studied based on the structures optimized by quantum mechanics methods or extracted from MD simulation. The results show consistency with our explicit and implicit solvent models.

3. Methodology

3.1. Explicit Solvent Model

The classical thermodynamics has its special superiority on the prediction of macroscopic equilibrium properties. Even so, it cannot be directly adopted to handle the nonequilibrium phenomena. Leontovich [53] once introduced a constrained equilibrium approach to treat nonequilibrium states and it can be recast for deriving the electrostatic solvation energy for the nonequilibrium polarization [21].

The constrained equilibrium approach is threefold in essence [21]. Firstly, by imposing suitable external conservative forces, a nonequilibrium state of an isothermal system without flow can always be mapped to a constrained equilibrium state meanwhile keeping the internal variables fixed. Secondly, the differences in state function between the so constructed constrained equilibrium and any other equilibrium state can be calculated simply by means of classical thermodynamics. Thirdly, the external forces can be removed suddenly without friction from the constrained equilibrium system so as to recover the true nonequilibrium situation. This approach was elaborated in Ref. [21].

In this part, the previous work in Ref. [21] in continuum model is extended to a new version thoroughly based on the explicit solvent model. For both the equilibrium and nonequilibrium polarizations, the total solvation energy is mainly composed of four parts [60, 61]: the cavitation formation energy, the dispersion and repulsion energy, the electrostatic energy as well as the solute distortion energy. In fact, when the Franck–Condon transition occurs in solution, the first two terms can be regarded unchanged while the electrostatic term varies a lot. The distortion energy is the energy cost to polarize the solute and it takes the form of [61]

$$U_1^{\text{eq, dist}} = \langle \psi_1 | \mathbf{H} | \psi_1 \rangle - \langle \psi_1^0 | \mathbf{H} | \psi_1^0 \rangle \tag{1}$$

$$U_2^{\text{non, dist}} = \langle \psi_2 | \mathbf{H} | \psi_2 \rangle - \langle \psi_2^0 | \mathbf{H} | \psi_2^0 \rangle \tag{2}$$

where the superscripts "eq" and "non" denote equilibrium and nonequilibrium states and the subscripts "1" and "2" denote the initial ground state and excited states, respectively. ψ_2^0 and

ψ_1^0 are the solute wave functions in vacuum, while ψ_2 and ψ_1 are in solution. $\boldsymbol{H}$ is the Hamiltonian in vacuum.

In the following we concentrate only on the derivation of the nonequilibrium electrostatic solvation energy including the solvent polarization in explicit solvent model.

In solution, the solvent should be regarded as the "system" and the others including the solute charge distribution are considered as the "ambient". Two physical quantities, electric field and solvent polarization, are adopted to represent each state, as shown in figure 3. The initial ground state $[\boldsymbol{E}_1^{\text{eq}}, \boldsymbol{P}_1^{\text{eq}}]$ will change to the nonequilibrium state $[\boldsymbol{E}_2^{\text{non}}, \boldsymbol{P}_2^{\text{non}}]$ with the solvent permanent dipoles fixed due to their slow response to the ultrafast variation of the solute charge distribution. Thus, we have

$$\sum_l \boldsymbol{\mu}_{2,l}^{\text{non,pmn}} = \sum_l \boldsymbol{\mu}_{1,l}^{\text{eq,pmn}} \tag{3}$$

$$\sum_j \boldsymbol{E}_{2,j}^{\text{non,pmn}}(\boldsymbol{r}) = \sum_j \boldsymbol{E}_{1,j}^{\text{eq,pmn}}(\boldsymbol{r}) \tag{4}$$

where the subscripts "*j*" and "*l*" indicate the number of the solvent molecule and the superscript "pmn" stands for the quantities related to the permanent dipole, respectively. At the nonequilibrium state, the solvent induced dipole has completed its fast response and reached the equilibrium with the new solute charge distribution. Furthermore, the solvent permanent dipoles will adjust the orientation and finally relax to equilibrate with the solute charge and the system will arrive at the equilibrium excited state $[\boldsymbol{E}_2^{\text{eq}}, \boldsymbol{P}_2^{\text{eq}}]$.

The constrained equilibrium state $[\boldsymbol{E}_2^{\text{non}} + \boldsymbol{E}_{\text{ex}}, \boldsymbol{P}_2^{\text{non}}]$ is constructed by applying an external field $\boldsymbol{E}_{\text{ex}}$ on the equilibrium excited state and finally this field will be removed suddenly to reproduce the nonequilibrium state, with the solvent total polarization fixed [21]. The new expression for the electrostatic solvation energy of nonequilibrium polarization is (see eqs. (25) and (30) and Appendix A in Ref. [21] for more information)

$$\Delta U_2^{\text{non}} = -\frac{1}{2}\int_V \boldsymbol{E}_2^{\text{non, vac}} \cdot \boldsymbol{P}_2^{\text{non}} \mathrm{d}V + \frac{1}{2}\int_V \boldsymbol{E}_{\text{ex}} \cdot \boldsymbol{P}_2^{\text{non}} \mathrm{d}V \tag{5}$$

where the superscript "vac" denotes the quantity due to the solute charges in vacuum.

For each state, the total electric field is composed of three parts: $\boldsymbol{E}^{\text{vac}}$ generated by the solute charges, $\boldsymbol{E}^{\text{pmn}}$ produced by the solvent permanent dipoles and $\boldsymbol{E}^{\text{ind}}$ due to the solvent induced dipoles. Thus the total electric field for each state at position $\boldsymbol{r}$ is

$$\boldsymbol{E}_1^{\text{eq}}(\boldsymbol{r}) = \sum_k \boldsymbol{E}_{1,k}^{\text{eq,vac}}(\boldsymbol{r}) + \sum_j \boldsymbol{E}_{1,j}^{\text{eq,pmn}}(\boldsymbol{r}) + \sum_j \boldsymbol{E}_{1,j}^{\text{eq,ind}}(\boldsymbol{r}) \tag{6}$$

$$\boldsymbol{E}_2^{\text{non}}(\boldsymbol{r}) = \sum_k \boldsymbol{E}_{2,k}^{\text{non,vac}}(\boldsymbol{r}) + \sum_j \boldsymbol{E}_{2,j}^{\text{non,pmn}}(\boldsymbol{r}) + \sum_j \boldsymbol{E}_{2,j}^{\text{non,ind}}(\boldsymbol{r}) \tag{7}$$

where the subscript "*k*" indicates the number of the solute atom. The total solvent polarization can be decomposed into two parts: the permanent dipole and the induced dipole, i.e.

$$\boldsymbol{P}_1^{\text{eq}} = \sum_l \boldsymbol{\mu}_{1,l}^{\text{eq,pmn}} + \sum_l \boldsymbol{\mu}_{1,l}^{\text{eq,ind}} \tag{8}$$

$$\boldsymbol{P}_2^{\text{non}} = \sum_l \boldsymbol{\mu}_{2,l}^{\text{non,pmn}} + \sum_l \boldsymbol{\mu}_{2,l}^{\text{non,ind}} \tag{9}$$

Given the solvent molecular polarizability tensor $\boldsymbol{\alpha}_l$, the induced dipole can be expressed by the total electric field as [59]

$$\boldsymbol{\mu}_{1,l}^{\text{eq,ind}} = \boldsymbol{\alpha}_l \cdot \left[\sum_k \boldsymbol{E}_{1,k}^{\text{eq,vac}}(\boldsymbol{r}_l) + \sum_j \boldsymbol{E}_{1,j}^{\text{eq,pmn}}(\boldsymbol{r}_l) + \sum_j \boldsymbol{E}_{1,j}^{\text{eq,ind}}(\boldsymbol{r}_l) \right] \tag{10}$$

$$\boldsymbol{\mu}_{2,l}^{\text{non,ind}} = \boldsymbol{\alpha}_l \cdot \left[\sum_k \boldsymbol{E}_{2,k}^{\text{non,vac}}(\boldsymbol{r}_l) + \sum_j \boldsymbol{E}_{2,j}^{\text{non,pmn}}(\boldsymbol{r}_l) + \sum_j \boldsymbol{E}_{2,j}^{\text{non,ind}}(\boldsymbol{r}_l) \right] \tag{11}$$

where the sum of the three terms in the parenthesis stands for the total field at the center of the *l*th solvent molecule.

In order to construct the constrained equilibrium state, an external field should be introduced. Actually, this field can be presented by the total field difference between the nonequilibrium state and the initial ground state (see eqs. (7) and (36) in Ref. [21] for more information). Accordingly, with combining eqs. (6), (7), and (4), the external field takes the form of

$$\boldsymbol{E}_{\text{ex}}(\boldsymbol{r}) = f\left[\boldsymbol{E}_2^{\text{non}}(\boldsymbol{r}) - \boldsymbol{E}_1^{\text{eq}}(\boldsymbol{r})\right] = f\left[\sum_k \boldsymbol{E}_{2,k}^{\text{non,vac}}(\boldsymbol{r}) + \sum_j \boldsymbol{E}_{2,j}^{\text{non,ind}}(\boldsymbol{r}) - \sum_k \boldsymbol{E}_{1,k}^{\text{eq,vac}}(\boldsymbol{r}) - \sum_j \boldsymbol{E}_{1,j}^{\text{eq,ind}}(\boldsymbol{r})\right] \tag{12}$$

with

$$f = \frac{\chi_{\text{op}}}{\chi_{\text{s}}} - 1 \tag{13}$$

here, χ_{op} and χ_{s} are the optical and static polarizabilities. With substituting eqs. (9) and (12) into eq. (5), the nonequilibrium electrostatic solvation energy in explicit solvent model can be achieved as

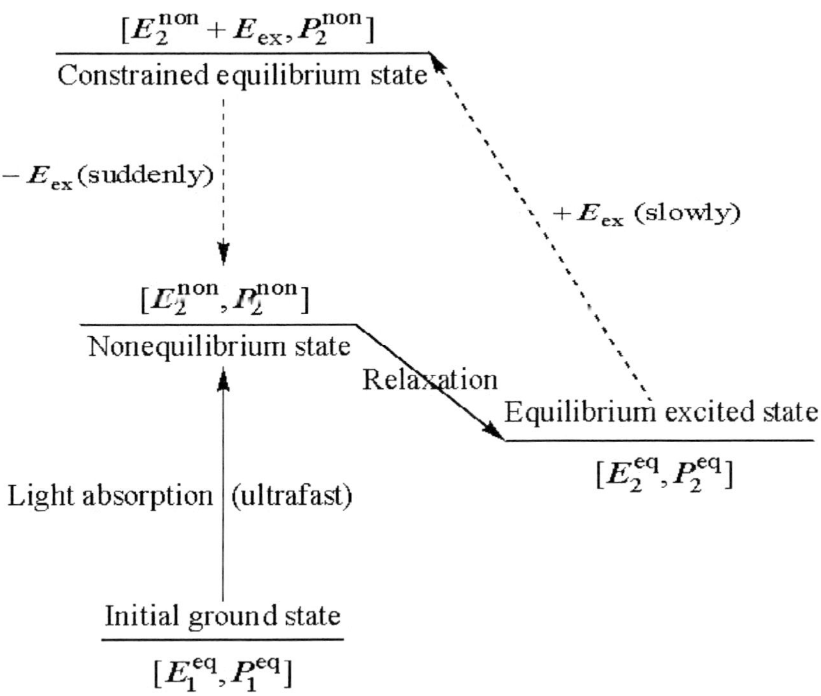

Figure 3. A generic schematic diagram for the constrained equilibrium approach. The system will be transformed into a nonequilibrium state from the initial ground state due to the ultrafast change of the solute charge distribution. It will further relax to a relatively stable state, equilibrium excited state. The constrained equilibrium state is constructed from the equilibrium excited state with adding an external field, which will be suddenly removed to recover the real nonequilibrium state.

$$\begin{aligned}\Delta U_2^{\mathrm{non}} = & -\frac{1}{2}\sum_l\sum_k \boldsymbol{E}_{2,k}^{\mathrm{nonvac}}(\boldsymbol{r}_l)\cdot(\boldsymbol{\mu}_{2,l}^{\mathrm{nonpmn}} + \boldsymbol{\mu}_{2,l}^{\mathrm{nonind}}) + \\ & \frac{f}{2}\sum_l\left[\sum_k \boldsymbol{E}_{2,k}^{\mathrm{nonvac}}(\boldsymbol{r}_l) + \sum_j \boldsymbol{E}_{2,j}^{\mathrm{nonind}}(\boldsymbol{r}_l) - \sum_k \boldsymbol{E}_{1,k}^{\mathrm{eq,vac}}(\boldsymbol{r}_l)\right. \\ & \left. -\sum_j \boldsymbol{E}_{1,j}^{\mathrm{eqind}}(\boldsymbol{r}_l)\right]\cdot(\boldsymbol{\mu}_{2,l}^{\mathrm{nonpmn}} + \boldsymbol{\mu}_{2,l}^{\mathrm{nonind}})\end{aligned} \tag{14}$$

If the solvent permanent dipoles are represented by the permanent charges, the interaction energy between the solute and solvent permanent charges can be expressed by the charge–potential form

$$\sum_l \boldsymbol{\mu}_l^{\mathrm{pmn}}\cdot\boldsymbol{E}^{\mathrm{vac}}(\boldsymbol{r}_l) = -\sum_i q_i V^{\mathrm{vac}}(\boldsymbol{r}_i) \tag{15}$$

where "i" denotes the solvent atom. The interaction between the permanent and induced dipoles can be equivalently expressed by

$$\sum_l \boldsymbol{\mu}_l^{\text{pmn}} \cdot \sum_j \boldsymbol{E}_j^{\text{ind}}(\boldsymbol{r}_l) = \sum_l \boldsymbol{\mu}_l^{\text{ind}} \cdot \sum_j \boldsymbol{E}_j^{\text{pmn}}(\boldsymbol{r}_l) \tag{16}$$

By combining eqs. (3), (15), and (16), eq. (14) can be rewritten as

$$\begin{aligned}\Delta U_2^{\text{non}} = & \frac{1-f}{2}\left[\sum_i\sum_k q_{2i}^{\text{non}} V_{2k}^{\text{non,vac}}(\boldsymbol{r}_i) - \sum_l\sum_k \boldsymbol{\mu}_{2l}^{\text{non,ind}} \cdot \boldsymbol{E}_{2k}^{\text{non,vac}}(\boldsymbol{r}_l)\right] \\ & + \frac{f}{2}\left[\sum_l\sum_j \boldsymbol{\mu}_{2l}^{\text{non,ind}} \cdot \boldsymbol{E}_{2,j}^{\text{non,pmn}}(\boldsymbol{r}_l) + \sum_l\sum_j \boldsymbol{\mu}_{2l}^{\text{non,ind}} \cdot \boldsymbol{E}_{2,j}^{\text{non,ind}}(\boldsymbol{r}_l) + \sum_i\sum_k q_{1i}^{\text{eq}} V_{1k}^{\text{eq,vac}}(\boldsymbol{r}_i)\right. \\ & \left. - \sum_l\sum_j \boldsymbol{\mu}_{1l}^{\text{eq,ind}} \cdot \boldsymbol{E}_{1,j}^{\text{eq,pmn}}(\boldsymbol{r}_l) - \sum_l\sum_k \boldsymbol{\mu}_{2l}^{\text{non,ind}} \cdot \boldsymbol{E}_{1k}^{\text{eq,vac}}(\boldsymbol{r}_l) - \sum_l\sum_j \boldsymbol{\mu}_{2l}^{\text{non,ind}} \cdot \boldsymbol{E}_{1,j}^{\text{eq,ind}}(\boldsymbol{r}_l)\right]\end{aligned} \tag{17}$$

Eq. (17) contains the contributions of the solute–solvent electrostatic interaction and the solvent polarization energy. If all the subscripts "2" in eq. (17) are replaced by "1", the electrostatic solvation energy of the initial ground state in explicit solvent model can be obtained as

$$\Delta U_1^{\text{eq}} = \frac{1}{2}\sum_i\sum_k q_{1,i}^{\text{eq}} V_{1,k}^{\text{eq,vac}}(\boldsymbol{r}_i) - \frac{1}{2}\sum_l\sum_k \boldsymbol{\mu}_{1,l}^{\text{eq,ind}} \cdot \boldsymbol{E}_{1,k}^{\text{eq,vac}}(\boldsymbol{r}_l) \tag{18}$$

This is consistent with the corresponding equilibrium solvation energy expression in continuum model [62]. As the equilibrium equation for the solvation energy is well established, it should be regarded as one criterion for the reliability of the nonequilibrium formula whether it can be converted to the equilibrium one. Taking the solute distortion into account, the absorption spectral shift Δhv, defined as the difference of the total solvation energy between the nonequilibrium state and the initial ground state, can be formulized as

$$\Delta hv = \Delta U_{2,\text{total}}^{\text{non}} - \Delta U_{1,\text{total}}^{\text{eq}} = \Delta U_2^{\text{non}} - \Delta U_1^{\text{eq}} + U_2^{\text{non,dist}} - U_1^{\text{eq,dist}} \tag{19}$$

In this work, we select the ASEP/MD program to implement our new expression of the electrostatic solvation energy for the nonequilibrium polarization by modifying the nonequilibrium module of the program. Since there are parameters of the initial ground state in eq. (17), more information such as electric potential, field strength and induced dipoles are conserved during the equilibrium polarization calculation and invoked at the nonequilibrium polarization calculation in M-ASEP/MD program. For the induced dipoles of the jth solvent molecule in eqs. (10) and (11) are the function of the other induced dipoles, the equations should be solved iteratively. As commented by Öhrn et al. [63] and Rösch et al. [64], the spectral shift from the dispersion/repulsion interaction can be ignored due to its nearly equal contribution to both the ground and excited states, for strong polar solvents. Thus, it appears that the ASEP/MD is one adaptive program to estimate the spectral shift in water solvent with sufficiently considering the mutual polarization and electrostatic interactions between the solute and solvent molecules. Notably, there is another type of nonequilibrium that is implicit in the QM/MM calculations even at the initial equilibrium ground state. More specifically, the solute–solvent structure is firstly obtained with a MD simulation. In this case, the solute is in equilibrium with the solvent structure. After a QM calculation for the solute, the solute properties, including the electronic and/or the geometrical structure, have changed with keeping the solvent surrounding fixed which is determined by the previous MD simulation.

Thus the previous equilibrium between the solute and solvent is disturbed and the nonequilibrium occurs. Then the solute–solvent will be simulated by MD again and the solute will be treated by the QM once more. Here the difference of the energy and charge population of the solute between these two calculations will be estimated and compared with the convergence criterion to determine whether the next iteration should be performed. Finally, when the convergence criterion is reached, the solvent structure is basically accounted in equilibrium with the solute. Therefore, this kind of nonequilibrium is different from the one due to the ultrafast light absorption, and this nonequilibrium effect cannot be completely eliminated in a sequential QM/MM approach, but can be controlled in a reasonable error range. It appears that the QMSTAT model that exploits a hybrid approach to treat the connection between the QM region and solvent region, is able to solve this problem. It formulates a combined quantum-classical intermolecular potential and then uses it to construct configurations with the MC simulation [65].

3.2. Implicit Solvent Model

On the basis of our previous derivation in continuum model [21], an analytical expression of the solvational shift of the spectra is deduced in this part with the approximation of the point dipole and sphere cavity. The solute charge distribution is considered as a point dipole, locating at the center of a single vacuum sphere with the radius of a. The cavity is surrounded by the solvent with a static dielectric constant of ε_s. The solute dipole will change from μ_1 to μ_2 due to the Franck–Condon transition in the ultrafast light absorption.

The electrostatic solvation energy of nonequilibrium polarization can be expressed as (see eqs. (30) in Ref. [21])

$$\Delta U_2^{\mathrm{non}} = -\frac{1}{2}\int_V \boldsymbol{E}_2^{\mathrm{vac}} \cdot \boldsymbol{P}_2^{\mathrm{eq}} \mathrm{d}V + \frac{1}{2}\int_V \boldsymbol{E}_{\mathrm{ex}} \cdot \boldsymbol{P}' \mathrm{d}V \tag{20}$$

which is equivalent to eq. (5). $\boldsymbol{P}'$ is called the residual polarization produced by the external vacuum field $\boldsymbol{E}_{\mathrm{ex}}$ in the medium and is defined as

$$\boldsymbol{P}' = \boldsymbol{P}_2^{\mathrm{non}} - \boldsymbol{P}_2^{\mathrm{eq}} \tag{21}$$

The external field $\boldsymbol{E}_{\mathrm{ex}}$ takes the form of (see eqs. (36) in Ref. [21])

$$\boldsymbol{E}_{\text{ex}} = \frac{\varepsilon_{\text{op}} - \varepsilon_{\text{s}}}{\varepsilon_{\text{s}} - 1} \Delta\boldsymbol{E}_{\text{op}} \tag{22}$$

with

$$\Delta\boldsymbol{E}_{\text{op}} = \boldsymbol{E}_2^{\text{non}} - \boldsymbol{E}_1^{\text{eq}} \tag{23}$$

where ε_{op} is the optical dielectric constant. The formula of the solvation energy for the initial ground state is uncontroversial and can be expressed as [21]

$$\Delta U_1^{\text{eq}} = -\frac{1}{2}\int_V \boldsymbol{E}_1^{\text{vac}} \cdot \boldsymbol{P}_1^{\text{eq}} \mathrm{d}V \tag{24}$$

The absorption spectral shift is defined as the difference of the total solvation energy between the nonequilibrium excited state and the initial equilibrium ground state; therefore we have

$$\begin{aligned} \Delta h\nu &= \Delta U_2^{\text{non}} - \Delta U_1^{\text{eq}} \\ &= \frac{1}{2}\int_V \boldsymbol{E}_{\text{ex}} \cdot \boldsymbol{P}'\mathrm{d}V - \frac{1}{2}\int_V \boldsymbol{E}_2^{\text{vac}} \cdot \boldsymbol{P}_2^{\text{eq}}\mathrm{d}V + \frac{1}{2}\int_V \boldsymbol{E}_1^{\text{vac}} \cdot \boldsymbol{P}_1^{\text{eq}}\mathrm{d}V \end{aligned} \tag{25}$$

Considering eq. (22) and the transformation expression of $\Delta\boldsymbol{E}_{\text{op}} = -\nabla(\Delta\varphi^{\text{op}})$, the above equation can be rewritten with the charge–potential representation as

$$\begin{aligned} \Delta h\nu = &\frac{1}{2}\frac{\varepsilon_{\text{op}} - \varepsilon_{\text{s}}}{\varepsilon_{\text{s}} - 1}\oint_S \Delta\varphi^{\text{op}}(\Delta\sigma^{\text{op}} - \Delta\sigma^{\text{eq}})\mathrm{d}S \\ &+ \frac{1}{2}\int_V \rho_2\varphi_2^{\text{eq}}\mathrm{d}V - \frac{1}{2}\int_V \rho_1\varphi_1^{\text{eq}}\mathrm{d}V \end{aligned} \tag{26}$$

where ρ_i and φ_i^{eq} (i=1 stands for the ground state, i=2 stands for the excited state) are the solute charge distribution and the polarization potential, respectively. $\Delta\varphi^{\text{op}}$ and $\Delta\sigma^{\text{op}}$ are the polarization potential and the surface polarized charge density generated by the change of the solute charge $\rho_2 - \rho_1$ in the medium with a dielectric constant of ε_{op}, respectively. $\Delta\sigma^{\text{eq}}$ is the surface polarized charge density generated by the change of the solute charge $\rho_2 - \rho_1$ in the medium with a dielectric constant of ε_{s}.

Since the solute charge distribution is treated as a point dipole located in a cavity with a radius of a, the polarization potential at position $\boldsymbol{r}$ and the surface polarized charge density can be described as

$$\varphi_i^{\text{op}} = \frac{3\mu_i \cos\theta}{(1 + 2\varepsilon_{\text{op}})r^2}, \quad i=1,2 \tag{27}$$

$$\sigma_i^{\mathrm{op}} = \frac{3(\varepsilon_{\mathrm{op}} - 1)\mu_i \cos\theta}{4\pi(2\varepsilon_{\mathrm{op}} + 1)a^3}, \quad i=1,2 \tag{28}$$

$$\sigma_i^{\mathrm{eq}} = \frac{3(\varepsilon_{\mathrm{s}} - 1)\mu_i \cos\theta}{4\pi(2\varepsilon_{\mathrm{s}} + 1)a^3}, \quad i=1,2 \tag{29}$$

where θ is the angle between $\boldsymbol{r}$ and the solute dipole. The dipole moment can be expressed by

$$\boldsymbol{\mu} = \lim_{\substack{\boldsymbol{L}\to 0 \\ q\to\infty}} q\boldsymbol{L} \tag{30}$$

where L is the distance pointing from q_+ to q_-. Thus, we have

$$\int_V \rho\varphi \mathrm{d}V = (q\varphi_+ - q\varphi_-) = \lim_{\substack{\boldsymbol{L}\to 0 \\ q\to\infty}} q\boldsymbol{L}\cdot\nabla\varphi = -\boldsymbol{\mu}\cdot\boldsymbol{E} \tag{31}$$

For the equilibrium solvation, we can obtain

$$\begin{aligned}\frac{1}{2}\int_V \rho_i\varphi_i^{\mathrm{eq}}\mathrm{d}V &= -\frac{1}{2}\boldsymbol{\mu}_i\cdot\boldsymbol{E}_i^{\mathrm{eq}} \\ &= -\frac{1}{2}\boldsymbol{\mu}_i\cdot\boldsymbol{\mu}_i\frac{2(\varepsilon_{\mathrm{s}} - 1)}{a^3(2\varepsilon_{\mathrm{s}} + 1)}, \; i=1,2 \\ &= -\frac{\mu_i^2(\varepsilon_{\mathrm{s}} - 1)}{a^3(2\varepsilon_{\mathrm{s}} + 1)}\end{aligned} \tag{32}$$

Combining eqs. (26)–(29) and (32) the final formula of the absorption spectral shift in the model of point dipole and sphere cavity can be achieved as

$$\Delta h\nu = \frac{9(\Delta\mu)^2(\varepsilon_{\mathrm{s}} - \varepsilon_{\mathrm{op}})^2}{a^3(2\varepsilon_{\mathrm{op}} + 1)^2(2\varepsilon_{\mathrm{s}} + 1)(\varepsilon_{\mathrm{s}} - 1)} + \frac{(\mu_1^2 - \mu_2^2)(\varepsilon_{\mathrm{s}} - 1)}{a^3(2\varepsilon_{\mathrm{s}} + 1)} \tag{33}$$

with

$$\Delta\mu = \boldsymbol{\mu}_1 - \boldsymbol{\mu}_2 \tag{34}$$

where $\boldsymbol{\mu}_1$ and $\boldsymbol{\mu}_2$ are the dipoles in solution. a is estimated with considering the solvent surrounding from the solute–solvent configurations extracted from the MD simulations. Eq. (33) is also applicable to the emission spectra in solution.

4. Computational Details

The geometry optimization of isolated formaldehyde is performed by the complete active space self-consistent field (CASSCF) [21] method with 8 active electrons in 8 active orbitals at the level of 6-31++G** with the Gaussian 98 program [66].

The optimized structure is shown in figure 4. Adopting the same calculation level, the vertical excitation calculations were carried out in vacuum and in aqueous solution with the M-ASEP/MD program. An approximation is made that the solute geometry is frozen when it is transferred from gas phase into solution. This may bring about some deviation for the cases in which the solute structure significantly depends on the solvent surrounding, but for formaldehyde, the approximation for the fixed solute structure has been found to be reliable with sufficient quality [41,57,65]. The Gaussian 03 packages [67] were also employed for the estimation of the spectral shift by the PCM model.

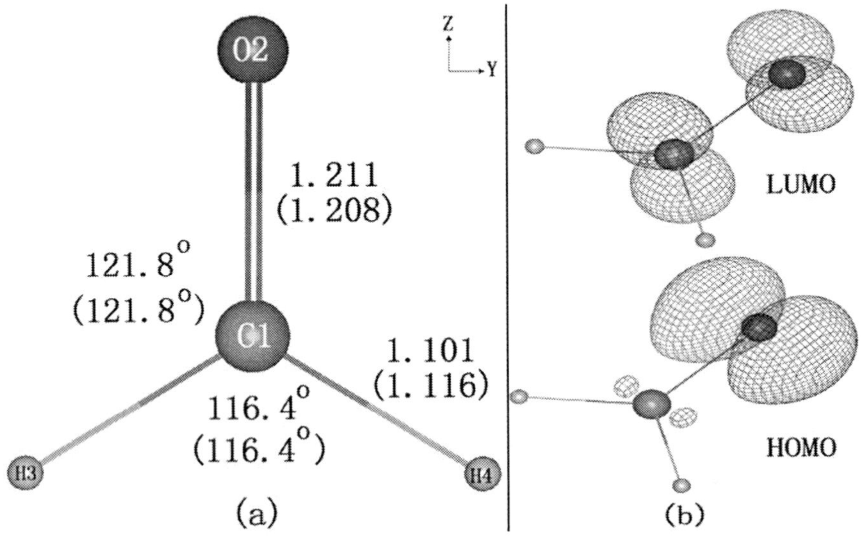

Figure 4. (a) The optimized geometry for FD with the CAS(8,8)/6-31++G** method. The data in parentheses are the experimental values. Bond length is in angstrom and bond angle in degree. (b) The HOMO and LUMO frontier orbitals.

Since in aqueous solution, formaldehyde tends to react with water to form a new substance methylendiol:

$$H_2CO + H_2O \rightleftharpoons H_2C(OH)_2$$

We make an estimation of the HOMO → LUMO transition for isolated methylendiol. The geometry optimization was performed with the CASSCF(8,8)/6-31++G** method and no imaginary frequency was found. Subsequently, the calculations were carried out to evaluate the excitation energy to the lowest covalent excited state, with employing the CASSCF(8,8)/6-31++G** and the TDDFT/6-31++G** method, respectively.

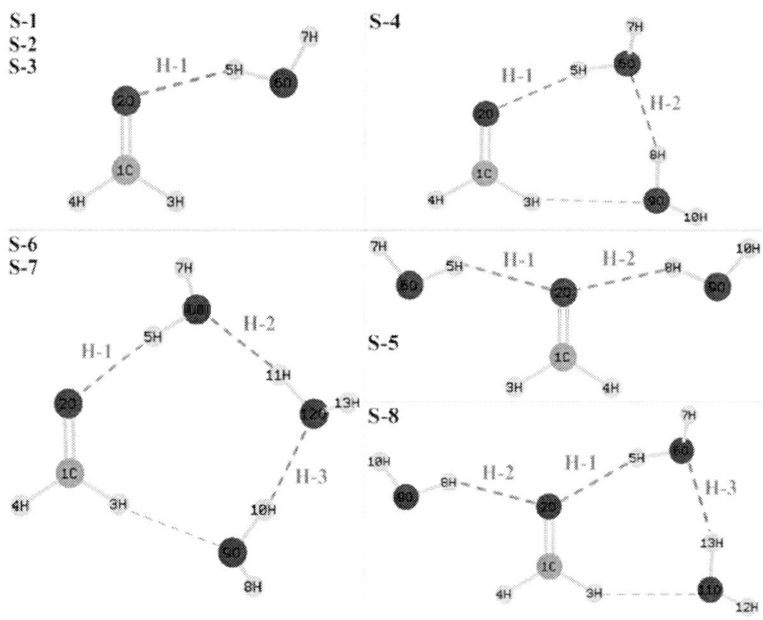

Figure 5. The structures for formaldehyde and water clusters optimized by CASSCF method. Three primary geometries are taken as the optimization starting point for $CH_2O–1H_2O$. The stationary points are more or less the same marked as S1, S2 and S3. Two optimal structures S4 and S5 are obtained for the two waters situation. Three optimal structures S6, S7 and S8 are found for the $CH_2O–3H_2O$ cluster. "H−1" denotes the nearest hydrogen bond between the formaldehyde carbonyl and water molecule. The "H−2" and "H−3" can be analogized in turn.

The MD simulations were carried out by the Moldy program [68]. The TIP3P water model [69] with fixed geometry was taken for the solvent molecules. In order to complement the deviation coming from such nonpolarizable treatment of the solvent, a polarization calculation is carried out sequentially with introducing the solvent molecular polarizability for considering the mutual polarization between the solute and solvent as well as the solvent induction contribution to the total spectral shift.

This is economical and has been proved to be equivalent to the full polarizable strategy. A cubic box with a side length of 18.7 Å was filled with 214 explicit water molecules and periodic boundary conditions were applied. A cut-off radius of 9 Å was set for the solute–solvent interaction. The temperature was fixed at 273 K using the Nosé–Poincaré thermostats [70], while the electrostatic interactions were handled using the Ewald Sum technique [71]. A total of 150000 steps were tracked with a steplength of 0.5 fs. The radial distribution functions (RDF) were calculated from 25000 fs and stored every 5 fs. The solute and solvent configuration data were dumped from 25000 fs with an interval of 500 fs and 100 configurations were taken into account to calculate the ASEP. In the case of non-polarizable model, the charges for a solvent molecule were set as $q(O) = –0.808$ and $q(H) = 0.404$, while in the polarizable model $q(O) = –0.716$ and $q(H) = 0.358$. These charges brought by the oxygen and hydrogen atoms and the solvent molecular polarizability were obtained by the Gaussian calculation with the method of HF/6-31++G**. We present a supermolecule model to study the hydrogen bonding effect on the lowest singlet n $\rightarrow$ π^* transition spectra of formaldehyde in aqueous solution. The structures of the clusters of $CH_2O–nH_2O$ (n=1, 2, 3) are optimized by CASSCF(8,8)/6-31++G** method. According to other theoretical studies about the formaldehyde–water clusters [30, 32, 33, 35, 46, 52, 72], three most probable structures are taken as the optimization starting point for CH_2O–$1H_2O$. Interestingly, these three parallel optimizations initiating from disparate geometries acquire a similar co-planar structure marked as S1, S2 and S3 in figure 5.

For the situation of two water molecules coordinated, there are two optimal structures S4 and S5 shown in figure 5. For CH_2O–$3H_2O$, three stationary points, S6, S7 and S8, were found without imaginary frequencies, in which structure S6 and S7 are similar. The following excitation calculations for these isolated clusters are carried out by both TDDFT and CASSCF(12,10) methods at 6-31++G** basis set. The carbonyl n, π and π* molecular orbits as well as the n orbitals located on water molecules are incorporated into the active space.

In our previous study of the hydrogen bonding effect on the absorption spectra of acetone [57], it is found that the supermolecule structures optimized by QM methods in vacuum are different from those extracted from simulation techniques. Thus, we analyzed the snapshots the MD simulation sampled and picked up representative CH_2O–nH_2O (n=1, 2, 3) structures (see S9, S10, S11, S12, S13 and S14 structures in figure 6) to perform TDDFT/6-31++G** calculations of excitation energies.

5. Results and Discussions

5.1. Molecular Properties

The optimized structure of isolated formaldehyde is shown in figure 4 with experimental data [73]. The CASSCF(8,8)/6-31++G** method seems good enough to obtain a reliable structure. For instance, the optimized C=O bond length, which has great effect on the n → π* transition as reported by [74, 75, 76], is only 0.003 Å larger than the experiment value [73]. The HOMO and LUMO frontier orbitals are also illustrated in figure 4. It is easy to recognize that the HOMO → LUMO excitation corresponds to one electron transfer from the lone pair n orbital of the oxygen to the π* orbital of the carbonyl group, which leads to some change of the dipole moment.

Some available theoretical results [3, 18, 31, 37, 40] and experimental observations [77, 78] are listed in table 1, and the dipole change due to the excitation is 0.70 Debye in vacuum, which agrees well with the experimental value of 0.74 Debye. For both the ground and excited states, our estimated dipole moments are also in good accord with other calculations. Different from the excitation that weakens the dipole, the solvent effect tends to strengthen it.

Such an enhancement is defined as a positive induced dipole of the solute. According to the data of the second and the fifth columns in table 1, it seems that, on one hand, the solvent effect makes the dipole moments increase ~50% both for ground and excited states, which is in line with the results of an equivalent 60% increase by Kawashima et al. [31]; on the other hand, the induced dipole at ground state is larger than that at excited state, and the larger the dipole moment is in vacuum, the larger the dipole moment will change when moving the solute from vacuum into solution. This implies a blue shift of the transition will occur.

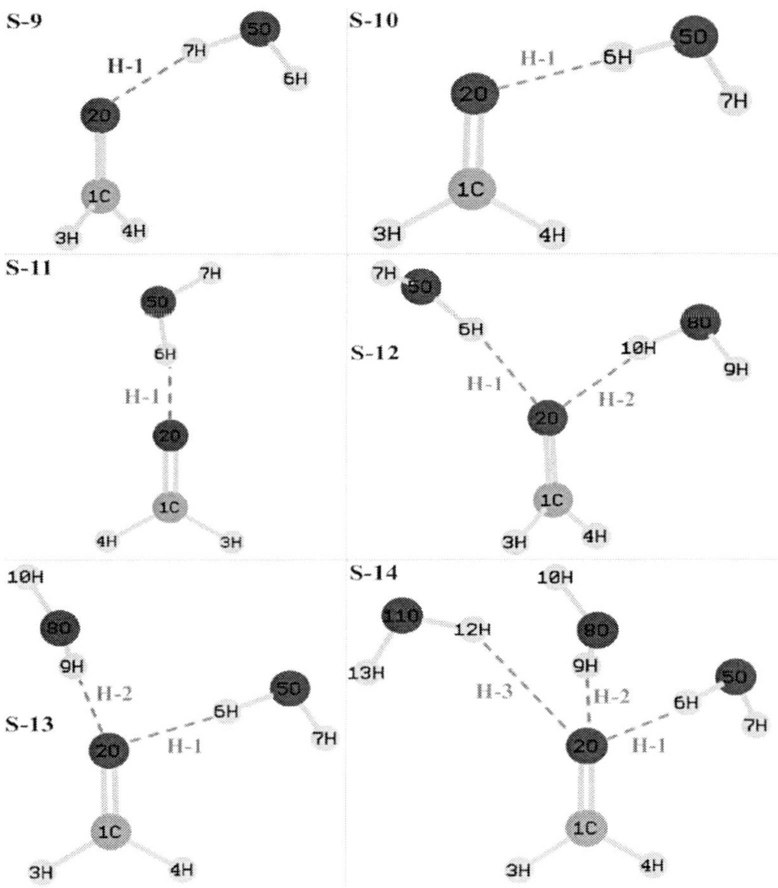

Figure 6. Representative structures of formaldehyde with neighbor waters extracted from MD trajectories. “H−1” denotes the nearest hydrogen bond between the formaldehyde carbonyl and water molecule. The “H−2” and “H−3” can be analogized in turn.

5.2. Vertical Excitation in Vacuum

The vertical excitation energy (VEE) of the n → π* transition of formaldehyde in vacuum is 4.01 eV with CASSCF(8,8)/6-31++G** in this work, which agrees well with the experimental 4.07 eV [79] with an error of 0.06 eV. With the more precise couple cluster method, Paterson et al. [36] employed the CCSD/aug-cc-pVDZ method to investigate the n → π* transition of formaldehyde in gas phase and got a VEE of 4.01 eV, not better than that obtained by the CASSCF method. Matsuzawa et al. [80] obtained a VEE of 4.00 eV with the TD-DFT(B3LYP)/DZP method, while Lin and co-workers [81] gained a VEE of 4.13 eV with the CASSCF(8,16)/6-311++G** method with adopting more active orbitals. It can be seen that the CASSCF method is appropriate and good enough to treat this kind of excitation with sufficient consideration of configuration interactions.

Table 1. Dipole moments for ground and excited states in vacuum and in water solution as well as induced dipole moments for ground and excited states due to the solvent effects (units in Debye)[a]

System	Dipole moments			Induced dipole moments	
	This work	Other calcs.	Expt.	This work	Other calcs.
Ground in vacuum	2.20	2.25[b], 2.11[c]	2.30[d]	1.11	0.65[f], 1.26[c]
Ground in water	3.31	3.34[e], 3.37[c]			
Excited in vacuum	1.50	1.47[g], 1.50[b]	1.56[h]	0.65	0.52[f], 1.01[c]
Excited in water	2.15	2.15[i], 2.34[c]			

[a] CAS(8,8)/6-31++G** method.
[b] MCSCF (MRCI)/cc-pVTZ (aug-cc-pVTZ) from Ref. [37].
[c] QM/MM-pol-vib/CAV model with the CASSCF method and TIP3P water model from Ref. [31].
[d] Ref. [77].
[e] RISM-SCF/DZP from Ref. [3].
[f] CAS(6,4)/ANO from Ref. [40].
[g] CAS(4,3)/6-31G** from Ref. [18].
[h] Ref. [78].
[i] ΔSCF/6-31G** from Ref. [18].

5.3. Solvent Structure

The O (formaldehyde)–O (water) and O (formaldehyde)–H (water) radial distribution functions (RDFs) are shown in figure 7 in dashed and solid line, respectively. The intensity and the range of the peaks reflect the extent of the mutual interaction between the solute and the solvent. The first peak appears at ~1.80 Å for the O–H RDF and ~2.8 Å for the O–O RDF, implying a strong hydrogen bond formed between the solute and solvent [75], which can be demonstrated by the microscopic structures of the system extracted from the equilibrium MD simulations, as shown in figure 6. Only several solvent molecules adjacent to the carbonyl oxygen are presented. The structure that has three and two hydrogen bonds formed between the carbonyl oxygen and the solvent hydrogen is dominant over the dumped configurations. This is in agreement with the results of Blair et al. [27], who determined an average of 2.6 waters in the first solvation shell. The second peaks are coincident and locate at ~4.30 Å, and the third peaks are at ~7.40 Å.

These indicate there are three solvation shells around the solute in the range of about 8 Å and the vicinity of the solute is well structured. The RDF profiles in this work are similar to other results [27, 34, 41, 47]. Notably, the shift of spectra is often attributed to hydrogen bonding effect. Kulkarni et al. have studied the solvent effect on the spectra of hydrogen-bonded formaldehyde–water clusters [82]. Canuto and Coutinho[47] estimated the contributions of 1–4 hydrogen bonds to the solvent shift of the n → π* transition of formaldehyde in water with MC simulation. Fonseca et al. investigated the hydrogen bond effect on the solvation properties of acetone in supercritical water with the S-MC/QM approach [83]. A similar study has been carried out by us about acetone and a conclusion was drawn that our model is adaptive to estimate the spectral shift in water solvent with sufficiently considering the polarization and electrostatic effect in the hydrogen bond [57]. A detailed discussion about the hydrogen bonding effect on formaldehyde will be presented in the following.

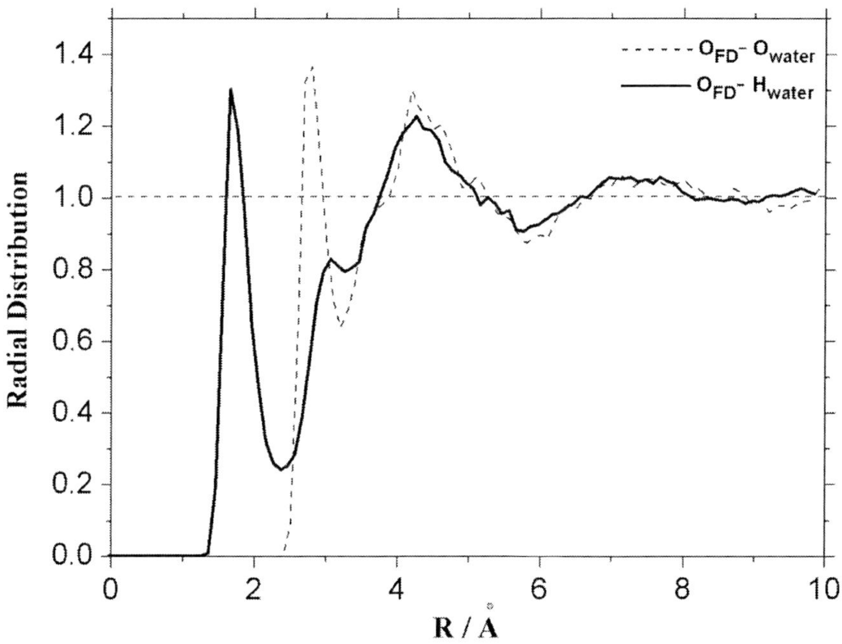

Figure 7. Oxygen (FD)–Oxygen (water) and Oxygen (FD)–Hydrogen (water) radial distribution functions of the last ASEP/MD cycle.

5.4. Solvent Shift Estimated with the M-ASEP/MD Program

The solvent shift of the n → π^* transition in formaldehyde in aqueous solution estimated with the M-ASEP/MD program is demonstrated in table 2, together with the result evaluated with PCM model. The three contributions, electrostatic, solvent polarization and the solute distortion, to the total solvent shift from the M-ASEP/MD calculations are also listed. A prediction of 0.20 eV is given by the M-ASEP/MD that employs eq. (19) and the CASSCF method, which is comparable to other theoretical results [24, 33, 35, 37, 39, 40, 41, 44, 47, 49, 51] reported in recent years with diverse models and methods, as shown in table 3.

The largest shift is 0.34 eV presented by Minezawa et al. [44] They applied the combined TDDFT and the polarizable effective fragment potential method to study the n → π^* vertical excitation of the hydrated formaldehyde. Another shift of 0.31 eV is given by Mochizuki et al. [51] with a supermolecular cluster including 16 waters with multilayer fragment molecular orbital model and parallelized integral-direct CIS(D) method. The other researchers got similar results to ours. With employing a discrete polarizable solvent representation and introducing the perturbation operator including the electrostatic and non-electrostatic terms, Öhrn and Karlström [41, 42] employed a hybrid approach to solve the connection between the

QM region and solvent region in their QMSTAT explicit solvation model. They exploited this program to study the spectral shifts of the n → π^* and π^* → n transitions in formaldehyde in aqueous solution. They received a solvent shift of 0.20 eV for the n → π^* absorption spectra and found the important contribution of the non-electrostatic interaction to the total shift.

Canuto and Coutinho [47] took three solvation shells that include eighty water molecules into account to estimate this solvent effect on the vertical excitation to the lowest excited state with their sequential QM/MC solvation model. They obtained a 0.24 eV of solvent shift. Although formaldehyde is frequently used as a test case to verify the rationality of models, it is a shame that the available experimental observations are insufficiently credible due to the fact that formaldehyde tends to form polymers and to react with water to bring about a new methylendiol.

This makes the direct comparison between a theoretical calculation and the experimental measurement difficult. Bercovici et al. [84] once carried out comprehensive spectral investigations for formaldehyde as a function of solvent and temperature. They detected a maximum absorption at 4.28 eV (290 nm) in water at room temperature at a ~10M formaldehyde concentration and attributed this absorption to the n → π^* transition of formaldehyde. Since formaldehyde in water is in equilibrium with methylendiol [41], this attribution may be inappropriate. Therefore, we estimated the HOMO → LUMO transition in isolated methylendiol with CASSCF(8,8)/6-31++G**, the same level of theory as that for formaldehyde. This gives an excitation energy of 7.74 eV (160 nm). Another TDDFT/6-31++G** calculation shows the excitation energy with a major contribution from the HOMO → LUMO transition is 8.74 eV (142 nm). It thus implies that the simple attribution of the 290 nm absorption to methylendiol is not reasonable. On the other hand, the inferences reported by some theoretical chemists [27, 31, 35, 39, 41] and the blue shift of acetone [85] imply that the solvent shift monomeric formaldehyde in aqueous solution is likely to be *ca.* 0.20 eV. Furthermore, the three contributions to the total solvent shift estimated by M-ASEP/MD are presented in table 2, including the electrostatic component from the interaction energy between the solute and solvent permanent charges, the polarization component from the interactions related to induced dipoles as well as the contribution from the difference of the solute's distortion energy between in nonequilibrium excited state and in initial ground state. The first two items are the difference between ΔU_2^{non} determined by eq. (17) and ΔU_1^{eq} determined by eq. (18).

Table 2. The solvatochromic shift estimated with M-ASEP/MD, including the electrostatic, solvent polarization and solute distortion contributions. The solvent shift evaluated with eq. (33) of our implicit solvent model is also listed (units in eV)

M-ASEP/MD[a]				Eq. (33)	PCM[b]	Expt.
Electrostatic	Polarization	Distortion	Total			
0.19	0.03	–0.02	0.20	0.17–0.20	0.13	0.21[i], 0.23[j]

[a] CASSCF(8,8)/ 6-31++G** method was employed in the M-ASEP/MD calculations.
[b] CAS(8,8)/6-31++G**/PCM with Gaussian03.

It is apparent that the first two components are the main contributions to the total solvent shift, of which the electrostatic contribution is the governing factor with a contribution more than 90%. The solvent polarization also plays a significant role and cannot be neglected. This finding is in accordance with the results of Minezawa et al. [44] and Slipchenko [45]. Minezawa et al. found the indirect component (consisting of the electrostatic and solvent polarization items) is the dominant factor in determining the excitation energies. Slipchenko reported that the electrostatic contribution dominates the solvatochromic shift and the polarization is responsible for about 20% of the shift. The solute's distortion is determined by eqs. (1) and (2) and is very small, less than that from the solvent polarization. Analogously, in Slipchenko's research [45], the distortion contributes ~5% to the total shift in the $n \rightarrow \pi^*$ transition in formaldehyde–water complexes with 2–6 water molecules. These give us a hint that our model can well handle this spectral shift for formaldehyde in water solution, since the electrostatic component including the solvent polarization is the main contribution to the solvent shift and it is what we concentrate on.

Usually, as its nearly equal contribution to both the states before and after the Franck–Condon transition in strong polar solvents, the spectral shift due to the dispersion/repulsion interaction is ignored in QM/MM strategies of evaluation of solvatochromic shift [63, 64], and so it is in this work. In this work, the solute geometry is fixed when it is transferred from the gas phase into solution. Lupieri et al. [39] found that the contribution from the geometric distortion of formaldehyde to the spectral shift can be ignored. Other contributions from the Stark effect and volume strain effect have been proved to be very small and can be neglected [22].

To conclude, since the permanent electrostatic and solvent polarization contributions control the spectra shift in formaldehyde in aqueous solution, our nonequilibrium polarization theory that focuses on the electrostatic component of solvation energy and our M-ASEP/MD program that iteratively solves the solvent polarization after a nonpolarizable MD simulation are reasonable and sufficient to consider the solvatochromic shift of formaldehyde in water.

5.5. Solvent Shift Estimated by the Implicit Solvent Model

To make estimation from eq. (33) with the point dipole and sphere cavity approximation, the cavity radius should be determined first. Behjatmanesh-Ardakani et al. [86] discussed the cavity shape affects the pK_a prediction of small amines with the PCM model. Improta and Barone [87] employed the PCM/TD-DFT method to study the cavity model effect on the $n \rightarrow \pi^*$and $\pi \rightarrow \pi^*$excitations of uracil derivatives in solution. Here, we propose a simple but reasonable scheme for the determination of the cavity radius. Taking the real solvent surrounding into account, the sphere radius is defined as the distance from the solute center to the nearest solvent atom from the MD simulations. More specifically, the amount of the solvent hydrogen atoms is taken as a function of the distance to the center of the solute molecule and averaged over the 100 dumped configurations.

Table 3. Comparison of our result with other theoretical predictions of the solvent shift in FD in aqueous solution with various models and methods (units in eV)

Authors	Model	Method	Shift
This work	M-ASEP/MD	CASSCF(8,8)/ 6-31++G**	0.20
Minezawa et al. [44]	TDDFT/EFP1	B3LYP/DZP	0.34
Lupieri et al. [39]	CPMD/MM, MD (TIP3P)	TDDFT-BLYP	0.23
Mochizuki et al. [51]	MLFMO, supermolecular cluster, 16 waters	CIS(D)/6-31G**	0.31
Xu and Matsika [37]	MRCI/MD(SPC)	MRCI/aug-cc-pVTZ	0.19
Öhrn and Karlström [41]	QMSTAT, MC(NEMO)	CASSI/ANO	0.20
Hirata et al. [49]	Binary interaction, 81 waters	EOM-CCSD/aug-cc-pVDZ	0.17
Kongsted et al. [35]	CC/MM	CCSD/aug-cc-pVTZ	0.24
Andrade do Monte et al. [24]	COSMO	MR-CISD/ANO	0.25
Dupuis et al. [33]	QM/MM-pol-vib/CAV, 81 waters (TIP3P)	CAS(12,10)/6-31G*	0.21
Martín et al. [40]	ASEP/MD	CAS(6,4)/ANO	0.18
Canuto and Coutinho [47]	QM/MM, MC(SPC), supermolecular cluster (80 waters)	INDO/CIS	0.24

The obtained profile is illustrated in figure 8. Three parts, corresponding to three solvation shells, can be identified, which is accord with that revealed in the RDFs in figure 7. The first shell is chiseled with the origin at 1.90 Å and the peak at 2.20–2.30 Å which is taken as the cavity radius. According to the dipoles in solution in table 1, eq. (33) predicts a shift of 0.17–0.20 eV in table 2, which agrees well with the experiments as well as that obtained by the M-ASEP/MD program. This suggests consistence of our theory in implicit and explicit solvent models and also gives us a hint that the electrostatic component including the induced polarizations is the main contribution to the blueshift.

It should be noticed that in our scheme the solvent effect on the solute radius has been considered with the aid of the MD simulation, which employs the Lennard-Jones potential to characterize the short–range dispersion/repulsion interactions and the Ewald Sum technique to handle the long–range electrostatic interactions. If the solvent effect on the solute radius is ignored, the traditional PCM with our optimized geometry gives a prediction of 0.13 eV, as shown in table 2.

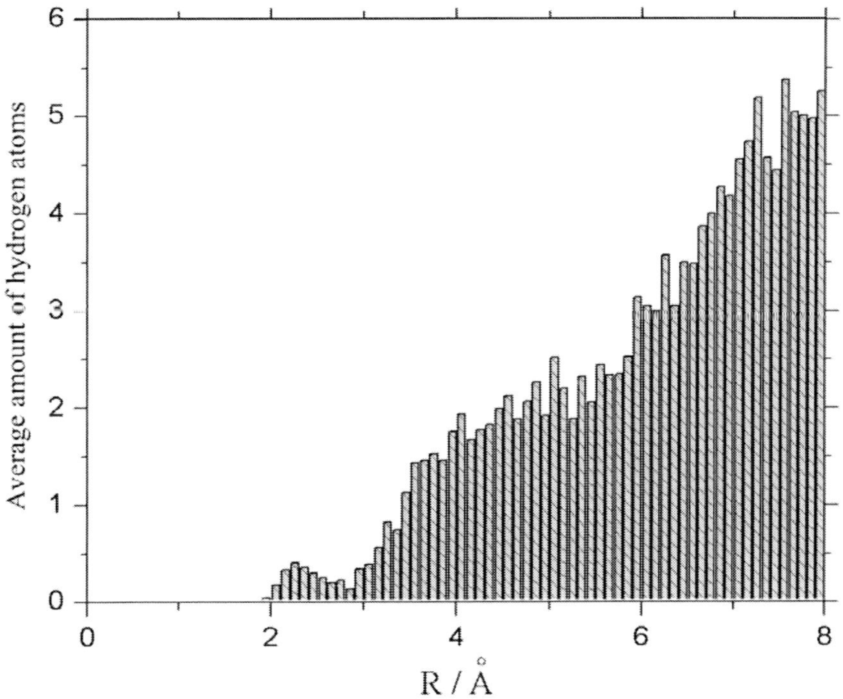

Figure 8. The profile for the statistical average distribution of the solvent hydrogen atoms according to the distance from the center of the solute molecule.

Comparing with the results obtained under explicit solvation strategies, one may conclude that the PCM model seems to underestimate the solvent shift of formaldehyde in water solution. Mennucci et al. [18] once employed the MRCI/6-31G**/PCM method to estimate the solvatochromic shift for the lowest-lying excitation of formaldehyde in water and got a shift of 0.12 eV. A similar situation has been found in the spectral shift in acetone, as detailedly discussed in our previous study [57]. This indicates that the determination of the cavity radius should take the strong specific solute–solvent interactions into account for polar solvents in continuum model. Canuto and Coutinho [47] reported the determination of the cavity radius should take not only the geometric criterion but also others like the energetic criterion into account. They found that it is quite appropriate to use the two criteria to identify the hydrogen bonds formed between the solute and the solvent molecules.

5.6. Solvent Shift Estimated by the Supermolecule Model

The solvatochromic shift of the n → π* transition is often attributed to hydrogen bonding effect, i.e. the hydrogen bond strength is different in the ground state and the excited state. In order to give a terse discussion of the hydrogen bond effect, the complexes of formaldehyde–H_2O, formaldehyde–$2H_2O$ and formaldehyde–$3H_2O$ were optimized in vacuum. The stable structures are shown in figure 5, and the related hydrogen bonding parameters are collected in tables 4 and 5 including the distance of accepter...hydrogen, donor...accepter and the angle of accepter...donor–hydrogen. "H–1" denotes the nearest hydrogen bond between the formaldehyde carbonyl and water molecule. The “H−2” and “H−3” can be analogized in turn. Although formaldehyde basically keeps a plane structure, the bond lengths and angles are more or less influenced by the coordination of water molecules.

The extent is quite coupled with the hydrogen bonding strength illustrated in table 4. Taking the structures S1, S2 and S3 for example, S1 and S3 have similar distance and angle parameters that imply a stronger hydrogen bonding effect than that of S2. Thus the bond length of C=O changed about 0.02 Å for S1 and S3 with respect to the isolated formaldehyde structure S0. In table 4, the distance 1.80 Å between the oxygen atom of formaldehyde and the nearest water hydrogen atom together with that 2.80 Å between the oxygen atom of formaldehyde and the nearest water oxygen in the CH2O−∞H2O cluster, i.e. the formaldehyde in the bulk water solvent, was directly obtained from the RDFs in figure 7.

Table 4. Hydrogen bonding parameters for the CH_2O–nH_2O (n=1, 2, 3) clusters (angstrom for distance and degree for angle) [a]

System	H–1			H–2			H–3		
	r(O...H)	r(O...O)	∠O...O–H	r (O...H)	r (O...O)	∠O...O–H	r (O...H)	r (O...O)	∠O...O–H
S1	2.11	2.97	20.7						
S2	2.17	3.02	23.9						
S3	2.11	2.97	21.2						
S4	2.05	2.97	12.7	1.99	2.90	6.4			
S5	2.17	2.99	24.4	2.17	2.99	24.4			
S6	2.01	2.96	4.8	3.5	1.93	2.9	1.95	2.90	3.05
S7	2.01	2.96	4.1	1.93	2.87	3.4	1.95	2.89	2.83
S8	2.07	2.98	13.5	2.14	2.98	22.5	1.99	2.90	7.0
S9	1.61	2.52	11.7						
S10	1.69	2.64	6.3						
S11	1.59	2.53	6.0						
S12	1.69	2.63	6.5	1.61	2.52	11.7			
S13	1.69	2.64	6.3	1.72	2.65	8.7			
S14	1.69	2.64	6.3	1.72	2.65	8.7	2.50	3.28	29.7
CH_2O–∞H_2O	1.80	2.80							

[a] S1–S14 are the cluster structures optimized by CASSCF or MD. r(3H–9O)=2.46 Å for S4; r(4H–9O)=2.80 Å for S5, r(3H–6O)=2.80 Å for S5;r(3H–9O)=2.33 Å for S6; r(3H–9O)=2.35 Å for S7; r(3H–11O)=2.42 Å for S8. “H-1” denotes the nearest hydrogen bond between the formaldehyde carbonyl and water molecule. The “H-2” and “H-3” can be analogized in turn. CH2O–∞H2O indicates the bulk solvent situation, of which the parameters are from figure 7.

It can be observed that the distances of r(O···H) and r(O···O) reduce step by step with increasing the number of the water molecules, and finally converge to the values in the bulk solvent with ASEP/MD simulations, which indicates that the hydrogen bond is strengthened when more solvent molecules are considered. It should be emphasized that the hydrogen bond formed in the complexes or clusters cannot stand for the hydrogen bond in the real liquid situation, when noticing the difference of the hydrogen bond lengths between in S1−S8 and in

S9-S14. Therefore, the meaning seems not significant when investigating the spectral shift of the complexes due to the hydrogen bond simply with the geometries optimized in vacuum. As emphasized by Desiraju [88], the hydrogen bond is not a simple interaction but complex conglomerate of four main and independent ingredients: [89] covalency (charge transfer), polarization, electrostatics, and van der Waals character (dispersion/repulsion).

Table 5. Parameters for isolated or clustered formaldehyde (angstrom for bond length and degree for angle)[a]

Parameter	S0	S1	S2	S3	S4	S5	S6	S7	S8
r(1-2)	1.21	1.23	1.21	1.23	1.23	1.21	1.21	1.23	1.21
r(1-3)	1.10	1.09	1.09	1.09	1.11	1.09	1.08	1.10	1.08
r(1-4)	1.10	1.09	1.09	1.09	1.11	1.09	1.09	1.11	1.09
∠213	121.8	121.0	121.2	121.0	121.3	121.1	121.7	121.7	120.8
∠214	121.8	120.8	120.9	120.7	120.0	121.1	120.0	119.8	120.0
Dihedral	180.0	180.0	180.0	180.0	180.0	180.0	179.9	179.9	179.9

[a] S0 is the isolated formaldehyde structure optimized by CASSCF. S1–S8 are the cluster structures optimized by CASSCF. r(1-2) denotes the bond length between atoms 1 and 2 of formaldehyde. Other parameters can be interpreted in similar way.

Traditionally, all hydrogen bonds were thought to be highly electrostatic and sometimes even partly covalent, and gradually the concept of a hydrogen bond became more relaxed to include weaker interactions, which have considerable dispersive repulsive character and merge into van der Waals interactions [88].

Our hydrogen bond changing from 2.17 Å to 1.69 Å in length in table 4 basically locates itself in the range of noncovalent interaction zone defined by Grabowski et al. [90] In ASEP/MD, the polarization contribution to the hydrogen bond can be estimated by the treatment of mutual polarization between the solute and the solvent with introducing the molecular polarizability. Therefore, the electrostatic interaction and weak interaction including the dispersion/repulsion are the main constituents of this moderate hydrogen bond, which is defined by Szatyłowicz [91] with hydrogen bond length in the range of 1.50−2.20 Å. As commented by Öhrn et al .[63] and Rösch et al., [64] the spectral shift due to the dispersion/repulsion interaction can be ignored due to its nearly equal contribution to both the ground and excited states, especially for strong polar solvents. Therefore, it seems that the ASEP/MD is one adaptive program to estimate the spectral shift in water solvent with sufficiently considering the polarization and electrostatic effect in hydrogen bond. The n → π* absorption spectral shift of formaldehyde due to the hydrogen bond effect was investigated by TDDFT and CASSCF methods for the $CH_2O–nH_2O$ (n=1, 2, 3) complexes with the structures optimized by CASSCF or MD. Results are collected in table 6.

For the structures from CASSCF, the excitation energy basically increases with adding water molecules. S2, with one water coordinated to formaldehyde, almost reproduces the shift, compared to the bulk solvent situation. $CH_2O–2H_2O$ gives the highest excitation energy. This is coupled with the information in table 5. S2 has a shorter C=O bond length with respect to that of S1 and S3 structures. When we take the CASSCF method to describe this system, with considering adequate configurations and reasonable components of the active space, S1 presents a 0.16 eV shift. CH_2O–$3H_2O$ gives smaller spectra shifts around 0.20 eV. The structures from MD have smaller shifts relative to those of the structures from CASSCF. A ~0.20 eV shift is expected for the hydrogen bonding effect on the n → π* absorption spectral of formaldehyde, which is consistent with our explicit and implicit solvent models.

Table 6. Vertical excitation energies and the corresponding solvatochromic shift for CH_2O–nH_2O (n=1, 2, 3) clusters (units in eV) [a]

System	VEE		Solvatochromic shift	
	TDDFT	CASSCF	TDDFT	CASSCF
S0	3.92	4.01		
S1	4.00	4.17	0.08	0.16
S2	4.10		0.18	
S3	4.00		0.08	
S4	3.99	4.32	0.07	0.31
S5	4.24		0.32	
S6	4.09		0.17	
S7	3.99		0.07	
S8	4.21		0.29	
S9	3.95		0.03	
S10	4.12		0.20	
S11	4.03		0.11	
S12	3.98		0.06	
S13	4.11		0.19	
S14	4.04		0.12	

[a] S0 is the isolated formaldehyde structure optimized by CASSCF. S1–S8 are the cluster structures optimized by CASSCF. S9–S14 are the cluster structures optimized by MD.

CONCLUSION

On the basis of our previous explorations, in the framework of continuous medium theory, the nonequilibrium polarization effect will occur in the solute ultrafast processes, e.g. electron transfer and photo-induced ionization and excitation, and an explicit solvation model is presented in this work to estimate the solvent effect on the UV/Vis absorption spectra in aqueous solution. Different from others' concerns, we concentrate on the establishment of proper electrostatic solvation energy formula for nonequilibrium polarization in explicit solvent model with adopting the constrained equilibrium approach. Unlike our early effort in explicit solvent model that emphasizes the physical image given by the continuum model, in this work, the formulation of the model was deduced entirely based on the explicit representation of the solvent surrounding, with considering the solvent at atomic-level and introducing discrete solvent dipoles and polarizability. The external field, applied to construct the virtual constrained equilibrium state, is expressed with the physical quantities in the initial equilibrium and final nonequilibrium state. The formulation was implemented under the popular QM/MD strategy with modifying the nonequilibrium module in the ASEP/MD program to carry out numerical calculations. We name the new procedure M-ASEP/MD that employs eq. (19) to evaluate the solvent shift. More information such as electric potential, field strength and induced dipoles are conserved during the equilibrium polarization calculation and invoked at the nonequilibrium polarization calculation in M-ASEP/MD program. The solvent polarization is iteratively solved after a nonpolarizable MD simulation. The new M-ASEP/MD codes were further applied to study the solvatochromic shift of the n $\rightarrow \pi^*$ transition in formaldehyde in aqueous solution.

The CASSCF(8,8)/6-31++G** method seems good enough to obtain a reliable structure and treat the vertical excitation of formaldehyde. The M-ASEP/MD program successfully reproduce the solvent structure of formaldehyde in water solution. A prediction of 0.20 eV is given by the M-ASEP/MD that employs eq. (19) and the CASSCF method, which is comparable to other theoretical results with diverse models and methods. The contribution from the electrostatic and solvent polarization is found to be dominant for the spectral shift, the same as concluded by others. This gives us a hint that our model can well handle this solvent shift for formaldehyde in aqueous solution, since the electrostatic component including the solvent polarization is the main contribution to the spectral shift and it is what we concentrate on. Due to the fact that formaldehyde in water is in equilibrium with methylendiol, we estimated the HOMO $\rightarrow$ LUMO transition in isolated methylendiol with CASSCF(8,8)/6-31++G** and TDDFT/6-31++G**. The results show that the absorption of methylendiol may be unlikely to mingle with the expected absorption of monomeric formaldehyde in aqueous solution.

A conjugated implicit solvent model is also proposed based on our energy expression of the nonequilibrium polarization with the approximation of the point dipole and sphere cavity. This model also gives good predictions of the solvatochromic shift of the n $\rightarrow \pi^*$ transition in formaldehyde in aqueous solution, showing self-consistency and reasonability of our theory. The supermolecule study of the CH_2O–nH_2O (n=1, 2, 3) clusters indicates that the electrostatic interactions are crucial for the hydrogen bonding effects between formaldehyde and water molecules. This type of hydrogen bonding can be well handled by our explicit solvent model and M-ASEP/MD program.

ACKNOWLEDGMENTS

This work is supported by the National Natural Science Foundation of China (Project No. 20903067) and the Scientific Research Foundation for the Returned Overseas Chinese Scholars, State Education Ministry (Project No. 20111139-10-10).

REFERENCES

[1] S.E. DeBolt, P.A. Kollman, *J. Am. Chem. Soc.* 112 (1990) 7515–7524.
[2] S. Ten-no, F. Hirata, S. Kato, *J. Chem. Phys.* 100 (1994) 7443–7453.
[3] K. Naka, A. Morita, S. Kato, *J. Chem. Phys.* 110 (1999) 3484–3492.

[4] N. Yoshida, S. Kato, *J. Chem. Phys.* 113 (2000) 4974–4984.
[5] Y. Mochizuki, S. Koikegami, S. Amari, K. Segawa, K. Kitaura, T. Nakano, *Chem. Phys. Lett.* 406 (2005) 283–288.
[6] Y. Ooshika, *J. Phys. Soc. Jpn.* 9 (1954) 594–602.
[7] R.A. Marcus, *J. Chem. Phys.* 24 (1956) 979–989.
[8] V.E. Lippert, Z. Naturforsch. 10a (1955) 541–545.
[9] N. Mataga, Y. Kaifu, M. Koizumi, *Bull. Chem. Soc. Jpn.* 29 (1956) 465–470.
[10] B.U. Felderhof, *J. Chem. Phys.* 1977, 67, 493.
[11] S. Lee, J.T. Hynes, *J. Chem. Phys.* 88 (1988) 6853–6862.
[12] H.J. Kim, *J. Chem. Phys.* 105 (1996) 6818–6832.
[13] M.A. Aguilar, F.J. Olivares del Valle, J. Tomasi, *J. Chem. Phys.* 98 (1993) 7375–7384.
[14] M.J. Powers, T.J. Meyer, *J. Am. Chem. Soc.* 1978, 100, 4393.
[15] B.S. Brunschwig, S. Ehrenson, N. Sutin, *J. Phys. Chem.* 1986, 90, 3657.
[16] M.D. Johnson, J.R. Miller, N.S. Green, G.L. Closs, *J. Phys. Chem.* 1989, 93, 1173.
[17] Onkelinx, F.C. De Schryver, L. Viaene, M. Van der Auweraer, K. Iwai, M. Yamamoto, M. Ichikawa, H. Masuhara, M. Maus, W. Rettig, *J. Am. Chem. Soc.* 1996, 118, 2892.
[18] Mennucci, R. Cammi, J. Tomasi, *J. Chem. Phys.* 1998, 109, 2798.
[19] V. Ganesan, S.V. Rosokha, J.K. Kochi, *J. Am. Chem. Soc.* 125 (2003) 2559–2571.
[20] F. Aquilante, V. Barone, B.O. Roos, *J. Chem. Phys.* 119 (2003) 12323–12334.
[21] X.Y. Li, F.C. He, K.X. Fu, W.J. Liu, *J. Theor. Comput. Chem.* 9, Supp. 1 (2010) 23–37.
[22] M.L. Sánchez, M.A. Aguilar, F.J. Olivares del Valle, *J. Phys. Chem.* 99 (1995) 15758–15764.
[23] K.V. Mikkelsen, A. Cesar, H. Ågren, H.J.A. Jensen, *J. Chem. Phys.* 103 (1995) 9010–9023.
[24] S. Andrade do Monte, T. Müller, M. Dallos, H. Lischka, M. Diedenhofen, A. Klamt, *Theor. Chem. Acc.* 111 (2004) 78–89.
[25] R. Improta, V. Barone, G. Scalmani, M.J. Frisch, *J. Chem. Phys.* 125 (2006) 054103.
[26] R. Improta, G. Scalmani, M.J. Frisch, V. Barone, *J. Chem. Phys.* 127 (2007) 074504.
[27] J.T. Blair, K. Krogh-Jespersen, R.M. Levy, *J. Am. Chem. Soc.* 111 (1989) 6948–6956.
[28] H. Fukunaga, K. Morokuma, *J. Phys. Chem.* 97 (1993) 59.
[29] Lie, G.C.; Clementi, E.; Yoshimine, *M. J. Chem. Phys.* 1976, 64, 2314.
[30] M.A. Thompson, J. Phys. Chem. 100 (1996) 14492.
[31] Y. Kawashima, M. Dupuis, K. Hirao, *J. Chem. Phys.* 117 (2002) 248–257.
[32] M. Dupuis, M. Aida, Y. Kawashima, K. Hirao, *J. Chem. Phys.* 117 (2002) 1242.
[33] M. Dupuis, M. Aida, Y. Kawashima, K. Hirao, *J. Chem. Phys.* 117 (2002) 1256.
[34] J. Kongsted, A. Osted, K.V. Mikkelsen, P.O. Astrand, O. Christiansen, *J. Chem. Phys.* 121 (2004) 8435–8445.
[35] J. Kongsted, A. Osted, T.B. Pedersen, K.V. Mikkelsen, *J. Phys. Chem. A* 108 (2004) 8624.
[36] M.J. Paterson, J. Kongsted, O. Christiansen, K.V. Mikkelsen, C.B. Nielsen, *J. Chem. Phys.* 125 (2006) 184501–184514.
[37] Z.R. Xu, S. Matsika, *J. Phys. Chem. A* 110 (2006) 12035–12043.
[38] T. Malaspina, K. Coutinho, S. Canuto, *J. Braz. Chem. Soc.* 19 (2008) 305.
[39] P. Lupieri, E. Ippoliti, P. Altoe, M. Garavelli, M. Mwalaba, P. Carloni, *J. Chem. Theory Comput.* 6 (2010) 3403.

[40] M.E. Martín, M.L. Sánchez, F.J. Olivares del Valle, M.A. Aguilar, *J. Chem. Phys.* 113 (2000) 6308–6315.
[41] Öhrn, G. Karlströmp, *Mol. Phys.* 104 (2006) 3087.
[42] Öhrn, G. Karlströmp, *J. Phys. Chem. A* 110 (2006) 1934.
[43] P. Arora, L.V. Slipchenko, S.P. Webb, A. DeFusco, M.S. Gordon, *J. Phys. Chem. A* 114 (2010) 6742.
[44] N. Minezawa, N.D. Silva, F. Zahariev, M.S. Gordon, *J. Chem. Phys.* 134 (2011) 054111.
[45] L.V. Slipchenko, *J. Phys. Chem. A* 114 (2010) 8824.
[46] Y. Dimitrova, S.D. Peyerimhoff, *J. Phys. Chem.* 97 (1993) 12731.
[47] S. Canuto, K. Coutinho, *Int. J. Quantum Chem.* 77 (2000) 192–198.
[48] K. Coutinho, S. Canuto, *J. Chem. Phys.* 113 (2000) 9132.
[49] S. Hirata, M. Valiev, M. Dupuis, S.S. Xantheas, S. SUGIKI, H. Sekino, *Mol. Phys.* 103 (2005) 2255.
[50] Y. Mochizuki, Y. Komeiji, T. Ishikawa, T. Nakano, H. Yamataka, *Chem. Phys. Lett.* 437 (2007) 66–72.
[51] Y. Mochizuki, K. Tanaka, K. Yamashita, T. Ishikawa, T. Nakano, S. Amari, K. Segawa, T. Murase, H. Tokiwa, M. Sakurai, Theor. *Chem. Acc.* 117 (2007) 541.
[52] M. Chiba, D.G. Fedorov, T. Nagata, K. Kitaura, *Chem. Phys. Lett.* 474 (2009) 227.
[53] M.A. Leontovich, *An Introduction to Thermodynamics,* 2nd ed., Gittl Publ, Moscow, 1950.
[54] X.Y. Li, Q.D. Wang, J.B. Wang, J.Y. Ma, K.X. Fu, F.C. He, *Phys. Chem. Chem. Phys.* 12 (2010) 1341–1350.
[55] H.Y. Wu, H.S Ren, Q. Zhu, X.Y. Li, *Phys. Chem. Chem. Phys.,* 14 (2012) 5538.
[56] X.J. Wang, Q. Zhu, Y.K. Li, X.M. Cheng, X.Y. Li, K.X. Fu, F.C. He, *J. Phys. Chem. B* 114 (2010) 2189–2197.
[57] Y.K. Li, Q. Zhu, X.Y. Li, K.X. Fu, X.J. Wang, X.M. Cheng, *J. Phys. Chem. A* 115 (2011) 232–243.
[58] Li, Y.K.; Wu, H.Y.; Zhu, Q.; Fu, K.X.; Li, X.Y. *Comput. Theoret. Chem.* 2011, 971: 65.
[59] I.F. Galván, M.L. Sánchez, M.E. Martín, F.J. Olivares del Valle, M.A. Aguilar, *Comput. Phys. Comm.* 155 (2003) 244–259.
[60] J. Tomasi, B. Mennucci, R. Cammi, Chem. Rev. 105 (2005) 2999–3093.
[61] M.E. Martín, A.M. Losa, I.F. Galván, M.A. Aguilar, *J. Chem. Phys.* 121 (2004) 3710–3716.
[62] J.D. Jackson, *Classical Electrodynamics*, 3rd ed., John Wiley and Sons Ltd, New York, 1999.
[63] Öhrn, G. Karlström, *Theor. Chem. Acc.* 117 (2007) 441–449.
[64] N. Rösch, M.C. Zerner, *J. Phys. Chem.* 98 (1994) 5817–5823.
[65] Öhrn, G. Karlström, in: S. Canuto (Ed.), Solvation Effects on Molecules and Biomolecules, *Springer Science+Business Media* B.V., 2008, pp. 215–246.
[66] M.J. Frisch, G.W. Trucks, H.B. Schlegel, G.E. Scuseria, M.A. Robb, J.R. Cheeseman, V.G. Zakrzewski, J.A., Jr. Montgomery, R.E. Stratmann, J.C. Burant, S. Dapprich, J.M. Millam, A.D. Daniels, K.N. Kudin, M.C. Strain, O. Farkas, J. Tomasi, V. Barone, M. Cossi, R. Cammi, B. Mennucci, C. Pomelli, C. Adamo, S. Clifford, J. Ochterski, G.A. Petersson, P.Y. Ayala, Q. Cui, K. Morokuma, D.K. Malick, A.D. Rabuck, K.

Raghavachari, J.B. Foresman, J. Cioslowski, J.V. Ortiz, A.G. Baboul, B.B. Stefanov, G. Liu, A. Liashenko, P. Piskorz, I. Komaromi, R. Gomperts, R.L. Martin, D.J. Fox, T. Keith, M.A. Al-Laham, C.Y. Peng, A. Nanayakkara, C. Gonzalez, M. Challacombe, P.M.W. Gill, B. Johnson, W. Chen, M.W. Wong, J.L. Andres, C. Gonzalez, M. Head-Gordon, E.S. Replogle, J.A. Pople, Gaussian 98, Revision A.11, Gaussian, Inc., Pittsburgh, 2001.

[67] M.J. Frisch, G.W. Trucks, H.B. Schlegel, G.E. Scuseria, M.A. Robb, J.R. Cheeseman, J.A. Montgomery, Jr., T. Vreven, K.N. Kudin, J.C. Burant, J.M. Millam, S.S. Iyengar, J. Tomasi, V. Barone, B. Mennucci, M. Cossi, G. Scalmani, N. Rega, G.A. Petersson, H. Nakatsuji, M. Hada, M. Ehara, K. Toyota, R. Fukuda, J. Hasegawa, M. Ishida, T. Nakajima, Y. Honda, O. Kitao, H. Nakai, M. Klene, X. Li, J.E. Knox, H.P. Hratchian, J.B. Cross, V. Bakken, C. Adamo, J. Jaramillo, R. Gomperts, R.E. Stratmann, O. Yazyev, A.J. Austin, R. Cammi, C. Pomelli, J.W. Ochterski, P.Y. Ayala, K. Morokuma, G.A. Voth, P. Salvador, J.J. Dannenberg, V.G. Zakrzewski, S. Dapprich, A.D. Daniels, M.C. Strain, O. Farkas, D.K. Malick, A.D. Rabuck, K. Raghavachari, J.B. Foresman, J.V. Ortiz, Q. Cui, A.G. Baboul, S. Clifford, J. Cioslowski, B.B. Stefanov, G. Liu, A. Liashenko, P. Piskorz, I. Komaromi, R.L. Martin, D.J. Fox, T. Keith, M.A. Al-Laham, C.Y. Peng, A. Nanayakkara, M. Challacombe, P.M.W. Gill, B. Johnson, W. Chen, M.W. Wong, C. Gonzalez, J.A. Pople, Gaussian 03, Revision D.01, Gaussian, Inc.: Wallingford CT, 2004.

[68] K. Refson, *Comput. Phys. Commun.* 126 (2000) 310–329.

[69] W.L. Jorgensen, J. Chandrasekhar, J.D. Madura, R.W. Impey, M.L. Klein, *J. Chem. Phys.* 79 (1983) 926–935.

[70] S. Nosé, *J. Phys. Soc. Jpn.* 70 (2001) 75–77.

[71] H.G. Petersen, *J. Chem. Phys.* 103 (1995) 3668–3679.

[72] R.A. Kumpf, J.R. Damewood, Jr., *J. Phys. Chem.* 93 (1989) 4478.

[73] C.M. Hadad, J.B. Foresman, K.B. Wiberg, *J. Phys. Chem.* 97 (1993) 4293–4312.

[74] K. Aidas, J. Kongsted, A. Osted, K.V. Mikkelsen, *J. Phys. Chem. A* 109 (2005) 8001–8010.

[75] O. Crescenzi, M. Pavone, F. De Angelis, V. Barone, *J. Phys. Chem. B* 109 (2005) 445–453.

[76] K. Coutinho, N. Saavedra, S. Canuto, *Theochem. – J. Mol. Struct.* 466 (1999) 69–75.

[77] J.W.C. Johns, A.R.W. McKellar, *J. Chem. Phys.* 63 (1975) 1682–1685.

[78] D.E. Freeman, W. Klemperer, *J. Chem. Phys.* 45 (1966) 52–57.

[79] M.B. Robin, *Higher excited states of polyatomic molecules*, Vol. III, Academic Press, New York, 1985.

[80] N.N. Matsuzawa, A. Ishitani, D.A. Dixon, T. Uda, *J. Phys. Chem. A* 105 (2001) 4953–4962.

[81] C.K. Lin, M.C. Li, M. Yamaki, M. Hayashi, S.H. Lin, *Phys. Chem. Chem. Phys.* 12 (2010) 11432–11444.

[82] A.D. Kulkarni, B. Mennucci, J. Tomasi, *J. Chem. Theory Comput.* 4 (2008) 578–585.

[83] T.L. Fonseca, K. Coutinho, S. Canuto, *Phys. Chem. Chem. Phys.* 12 (2010) 6660–6665.

[84] T. Bercovici, J. King, R.S. Becker, *J. Chem. Phys.* 56 (1972) 3956–3963.

[85] Renge, *J. Phys. Chem. A* 113 (2009) 10678–10686.

[86] R. Behjatmanesh-Ardakani, M.A. Karimi, A. Ebady, *Theochem. – J. Mol. Struct.* 910 (2009) 99–103.

[87] R. Improta, V. Barone, *Theochem. – J. Mol. Struct.* 914 (2009) 87–93.

[88] Desiraju, G. R. *Acc. Chem. Res.* 2002, 35, 565.

[89] Umeyama, H.; Morokuma, *K. J. Am. Chem. Soc.* 1977, 99, 1316.

[90] Grabowski, S. J.; Sokalski, W. A.; Dyguda, E.; Leszczyński, *J. J. Phys. Chem. B* 2006, 110, 6444.

[91] Szatylowicz, *H. J. Phys. Org. Chem.* 2008, 21, 897.

In: Formaldehyde
Editors: Chan Bao Cheng and Feng Hu Ln
ISBN 978-1-62257-214-4
©2012 Nova Science Publishers, Inc.

Chapter 4

DECONTAMINATION OF INDOOR AIR POLLUTANT OF FORMALDEHYDE THROUGH CATALYTIC OXIDATION OVER OXIDE SUPPORTED NOBLE METAL NANOCATALYSTS

Changyan Li[1,2], Baocang Liu[1,2], Yang Liu[1], Wenting Hu[1], Qin Wang[1] and Jun Zhang[1,2,*]

[1]College of Chemistry and Chemical Engineering, Inner Mongolia University, Hohhot, PR China

[2]College of Life Science, Inner Mongolia University, Hohhot, PR China

ABSTRACT

Formaldehyde (HCHO) is an important chemical feedstock and constituent of many industrial products, and is widely used in various adhesives and coatings of building materials. However, apart from its important application in chemical industry, HCHO is also defined as the most common and the best-known indoor air pollutant. Long time exposure to the indoor air with heavy HCHO pollution may cause serious health problems, such as irritation of the eyes, skin irritation, respiratory diseases, and even nasopharyngeal cancers. Thus, indoor air pollution has already aroused increasing concern, and great efforts have been made to eliminate HCHO pollution. Low temperature catalytic oxidation of HCHO is regarded as one of most attractive approach for elimination of HCHO, as HCHO can be completely converted into CO_2 and H_2O through catalytic oxidation process.

This paper is intended to review the recent advances in elimination of indoor air pollutant of HCHO through catalytic oxidation over oxide supported noble metal nanocatalysts. It is composed of four sections: (1) Overview of indoor air pollutant of HCHO including its chemistry, toxicology and source; (2) Various methods for elimination of HCHO pollution; (3) Decontamination of HCHO pollution through catalytic oxidation over oxide supported noble metal nanocatalysts; (4) Summary and outlook.

* Corresponding Author: Prof. Dr. Jun Zhang, Tel.: 0086 471 4992175; E-mail: cejzhang@imu.edu.cn.

Keywords: Indoor air pollutant, HCHO, Catalytic oxidation, Oxide supported noble metal nanocatalysts

1. INTRODUCTION

Formaldehyde (HCHO) was first described in the year 1855 by Alexander Michailowitsch Butlerow, a Russian scientist. August wilhelm von Hofmann, a German chemist, produced the HCHO by oxidation of methane or methanol in the presence of a catalyst in 1867 [1] Between 1900 and 1930, HCHO-based resins became important adhesives for wood and wood composites. In addition, HCHO also is an important chemical feedstock and constituent of many industrial products, which is widely used in various coatings of building materials, preservative, disinfectant, and biocide. Because long time exposure to the indoor air with heavy HCHO pollution may cause health problems, such as irritation of the eyes, skin irritation, respiratory diseases, and even nasopharyngeal cancer, HCHO is determined to be carcinogenic and teratogenic substance by the World Health Organization. It is recognized as a metamorphosis source and a major health killer in the daily life.

2. OVERVIEW OF INDOOR AIR POLLUTANT OF HCHO

2.1. Physical and Chemical Properties

HCHO is also called ant aldehydes. At room temperature, it is a colorless and flammable gas with irritant flavor. The size of molecular is approximately 0.3 nm and relative molecular weight is 30.03. In the air, the relative density of HCHO is 1.067, while the density of liquid comparing with water is 0.815 at room temperature 20°C. Melting and boiling points are respectively -92 °C and -21 °C. [2] HCHO is soluble in water, ethanol, diethyl ether, and acetone. It is commonly purchased as a 37% solution in water, known as formalin, with 10% methanol as a stabilizer. High purity HCHO could polymerize para HCHO with some catalysts. In addition, chemical property of HCHO is very active, which is easy to oxide and polymerize. Adhesives such as urea-HCHO (UF), melamine-urea-HCHO (MUF) and phenol-HCHO (PF) are usually synthesized according to equations of 1, 2, 3 and 4 respectively (Figure 1).[1]

2.2. The Sources and Hazards of HCHO

In indoor air, HCHO mainly comes from manual manufacturing and housing decoration, which is emitted from composite wood products (building materials and furniture), fiberglass insulation, paper products, fabrics including clothing and drapes and cosmetics. It is also emitted as a combustion product from gas and solid-fuel sources and tobacco smoke.

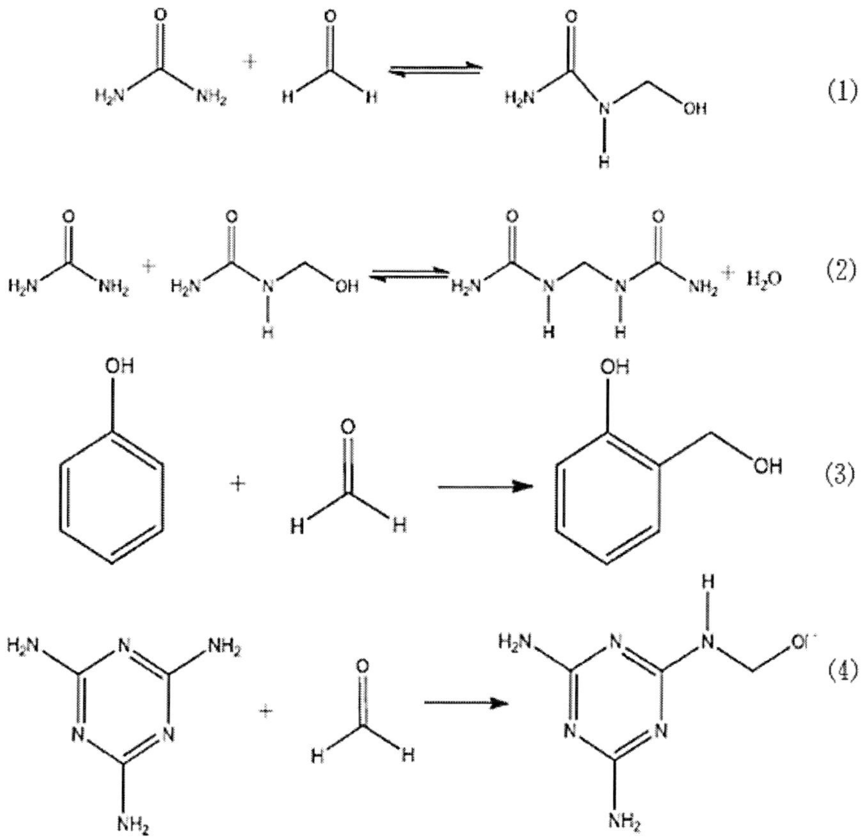

Figure 1. Reaction equations for formation of urea-HCHO (UF), melamine-urea-HCHO (MUF) and phenol-HCHO (PF).

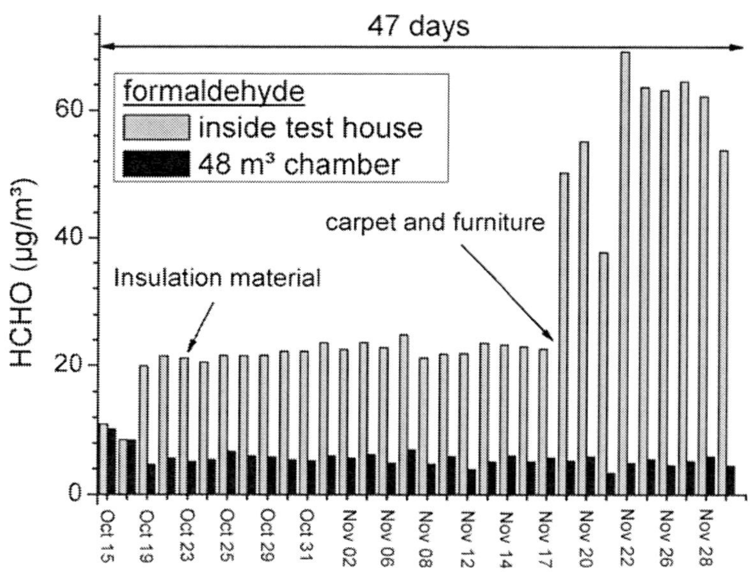

Figure 2. Influence of building products on the HCHO concentration inside a test house (n =0.3 h^{-1}) and in the 48 m^3 stainless steel chamber (n =2.0 h^{-1}) [1].

In particular, wood products fabricated with urea-HCHO resin such as particle board and medium-density fiberboard are the highest emitting persistent sources. These materials are used in large quantity in most new house construction. HCHO emissions from such sources are expected to persist over relatively long periods. Between 1996 and 2006, Germany Wihelm-Klauditz-Institue (WKI) Fraunhofer carried out 367 HCHO measurements in new prefabricated houses, which made with wood-based materials such as particle board and Oriented Strandboard (OSB). Survey results showed that fourteen percent data exceeded the German guidenline value of 0.1 ppm. It is well known that mineral wool often is installed in modern building product. In fact, mineral wool is also a major source of HCHO in the house. WKI once investigated the influence of the modern building products between Oct 15 and Nov 29 in 2007. The results of survey are shown in Figure 2. [1] Comparing with the test house (no mineral wool, carpeting, adhesives, or furniture) with 48 m^3 stainless steel chamber, it was learned that low concentrations of HCHO can hardly be avoided in a new living space. It is interesting that the concentration of HCHO in 48 m^3 stainless steel chamber is basically unchanged, while the presence of mineral wool had no influence on the HCHO level in the house (approximately 22 $\mu g/m^3$), but when the test house contained carpeting, carpet adhesive, and a sideboard made of lacquered particle board, an increase of the HCHO concentration up to 69 $\mu g/m^3$. On the other hand, the concentration of HCHO is related to the season. The HCHO concentration was 10 $\mu g/m^3$ in the summer and 100 $\mu g/m^3$ in the fall. The study of the new house has shown that HCHO is emitted by materials at relatively constant rates over a period of at least 9 months or decades. Park and Ikeda measured HCHO levels over a period of three years in new and older homes.[3] In the new homes, there was a falling trend from 134 μg/m3 (mean, first year) to 86 μg/m3 (mean, third year), while in the older homes the mean concentration of 88-90 μg/m3 became stable within three years. In a word, HCHO in indoor air has a close relationship with the environment.

HCHO has high solubility in water, which makes human body rapidly absorb in the respiratory and gastrointestinal tract. Here, it can be oxidized to form formate and exhaled as carbon dioxide. The biological half-life of HCHO is extremely short at about 1 min. [4] As an electrophile, HCHO can react with nucleophilic biogenic compounds in the body. [5] HCHO itself is produced in small amounts from methanol via the enzyme alcohol dehydrogenase (ADH), [6] which is a human metabolite and can be measured in urine. [7]

Table 1. Acute health effects from HCHO exposure [8]

HCHO concentration (ppm)	Observed health effects
< 0.05	None reported
0.05-1.5	Neurophysiologic effects
0.05-1.0	Odour threshold limit
0.01-2.0	Irritation of eyes
0.10-25	Irritation of upper airway
5-30	Irritation of lower airway and pulmonary e!ects
50-100	Pulmonary edema, inflammation, pneumonia
> 100	Coma, death

HCHO as a protoplasmic poison matter could couple with protein, so acute (short time) exposure to HCHO causes irritation of the eyes and the upper airways, and long-term exposure to lower levels has been associated with the increased risk of respiratory edema. Canada's federal department of health recently revised its residential indoor air quality guideline for HCHO: 1 and 8 h exposure limits were set at 123 and 50 mg/m^3, respectively. [7b] Adverse health effects from HCHO exposure may arise from inhalation or direct contact. A range of acute health impacts have been attributed to the substance (Table 1). [8]

In general, exposure to concentrations of less than 1 ppm may result in sneezing, coughing and minor eye irritation, and these symptoms often occur rapidly. But often absorbing HCHO, even if very low level, it will cause atopic dermatitis and color mottle, and sometimes it also induce asthma. Except dizziness, headache, body weakness, feeling of nausea, vomiting, chest tightness, or even a sore throat and so on symptom, HCHO also could make human weight loss, memory go down and even nervous disorder. In all population exposure to HCHO, children and pregnant woman are most sensitive to HCHO. Recently, HCHO has been classified as a human carcinogen by the International Agency for Research on Cancer. Some studies showed that HCHO is the culprit of nasopharyngeal cancer and leukemia. It is very necessary to eliminate HCHO indoors or in a closed environment.

3. Elimination Technology of HCHO in Indoor Air

Great efforts have been made to reduce the indoor concentration of HCHO to meet the stringent environmental regulations due to the increasing concern for human health. A number of methods have been proposed for elimination of HCHO, including physical adsorption, [9] plasma technology, [10] plant absorption, [11] photocatalysis [12] and catalytic oxidation. [13] HCHO can be oxidized into CO_2 and H_2O by catalytic oxidation at low cost using simple technology, and catalytic oxidation method could completely deal with much waste gas, showing some advantages including high efficiency, no adsorption saturation and secondary pollution. Thus, the catalytic oxidation is considered as one of the most important and promising technologies for HCHO elimination. Catalytic oxidation mainly includes two classes, namely photocatalytic oxidation and thermcatalytic oxidation.

3.1. Photocatalytic Oxidation

Photocatalysis has been demonstrated as an efficient abatement technology for catalytic oxidation of HCHO at ambient temperature and pressure conditions. Among the studied photocatalysts, titanium dioxide (TiO_2) is considered to be relatively inexpensive and chemically stable; therefore it has been extensively studied in UV-induced photocatalytic reactions. The major oxidative and reductive processes in the photocatalytic degradation of HCHO can be written as shown in Figure 3. [14] The UV irradiation activates TiO_2 to generate strongly oxidative holes (h_{VB}^+) in valence band and reductive electrons (e_{CB}^-) in conduction band. The h_{VB}^+ can oxidize the HCHO directly or react with H_2O to generate a free radical $^{\bullet}OH$ and H^+, and H^+ subsequently react with the absorbed O_2 to yield $^{\bullet}OH$. The free radical ${}^{\bullet}OH$ could also oxidize HCHO to form formic acid and further oxidize to CO_2 and H_2O.[15]

HCHO in indoor air can be eliminated by supporting TiO_2 on activated carbon fiber or other supports, which are made in power or thin film. These TiO_2 supported catalysts are placed at light reactor. Depending on adsorption and enrichment capabilities of porous materials, HCHO is oxidized under illumination. Some literature reported that TiO_2 supported on activated carbon fiber could adsorb and catalytically oxidize HCHO at 254 nm wavelength of ultraviolet light. Moreover, approximately 96% of HCHO could be eliminated. However, photocatalytic degradation of HCHO has some disadvantages, for instance, light source is very expensive and short life, and some poisonous substance could form during photocatalytic degradation of HCHO. Low concentration HCHO (< 1ppm) is difficult to eliminate and need very harsh conditions. The biggest problem in photocatalytic field is how to solve migration of electrons at visible light through changing photocatalyst structure.

3.2. Thermal Catalytic Oxidation

Thermal catalytic oxidation mainly uses supported metal or metal oxides catalysts to decompose or eliminate HCHO by physical and chemical method. According to reaction temperature, there are two kinds of thermal catalytic oxidation, namely, low and high temperature heterogeneous catalytic oxidation. The results of modern surface science show that oxygen adsorb on the transition metal and noble metal surface, which could dissociate adsorbed oxygen atoms or charged peroxy, superoxide radicals. These activated oxygen species are very easy to make HCHO turn into CO_2 and H_2O. HCHO catalytic oxidation technology carried out at room temperature may become hot topic in the field of indoor HCHO elimination. This paper will mainly review low temperature catalytic oxidation on the kinetics, mechanism and catalytic activity of supported noble metal nanocatalysts.

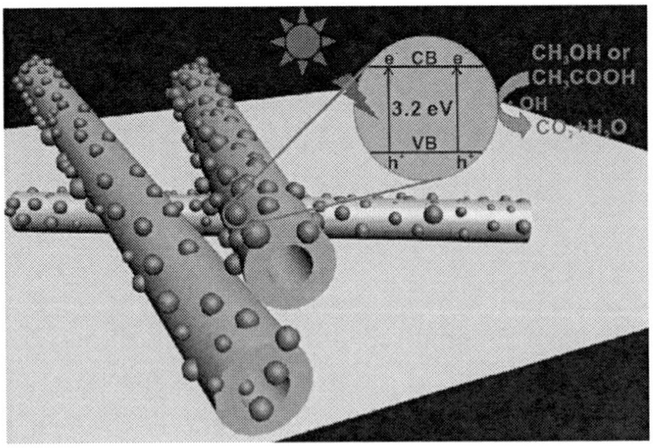

$$TiO_2 \xrightarrow{hv} e^- + h^+$$
$$HCHO + {}^{\bullet}OH \rightarrow {}^{\bullet}CHO + H_2O$$
$${}^{\bullet}CHO + {}^{\bullet}OH \rightarrow HCOOH$$
$${}^{\bullet}CHO + {}^{\bullet}O_2^- \rightarrow HCO_3^- \xrightarrow{+H^+} HCOOOH \xrightarrow{+HCHO} HCOOH$$
$$HCOOH \xrightarrow{-H^+} HCOO^- \xrightarrow{OH} H_2O + {}^{\bullet}CO_2^-$$
$$HCOO^- \xrightarrow{h^+} H^+ + {}^{\bullet}CO_2^-$$
$${}^{\bullet}CO_2^- \xrightarrow{[O],[{}^{\bullet}OH],h^+} CO_2$$

Figure 3. Photocatalytic process over TiO_2 (Left) and mechanism of HCHO photocatalytic oxidation (Right) [14-15].

4. HCHO Catalytic Oxidation over Oxide Supported Noble Metal Nanocatalysts

Catalytic oxidation is an attractive way to eliminate HCHO emissions to meet the requirement of air pollution control regulations. Supported noble metals have been found to show the good activity for catalytic oxidation of HCHO. Catalytic oxidation of HCHO most select noble metal such as Ru, Pd, Rh, Pt and Ir. [16] The activity of the catalysts was evaluated by the temperatures at which the conversion of HCHO reached 50% (T_{50}) and 90% (T_{90}). Ru/CeO_2 had the highest activity among the precious metals on CeO_2 investigated (Table 1). Later, Christoskova and coworkers prepared a kind of high oxidation activity Nickel oxide catalyst, [17] which could convert HCHO into CO_2 and H_2O at room temperature, but it was soon lose activity. Air pollution control regulations require this kind of catalyst possessing high activity at a low temperature. Of course, it is a big challenge to research workers in this area. In fact, low temperature catalytic oxidation of HCHO facing the most difficult problems is deactivation, catalytic low efficiency and high cost, which make thermal catalytic oxidation of HCHO difficult to industrialize. In the past twenty years, HCHO catalytic oxidation always has been focused on activity species, supports, mechanism and kinetics.

4.1. Factor of Influencing HCHO Catalytic Oxidation Activity

Considerable studies have been directed toward elucidating the effectiveness of noble metal catalysts deposited on commonly used supports such as alumina, silica, and silica-alumina. [18] Moreover, people initially thought that the role of the carrier is only carrying and dispersing active component, so most inert material was selected as carrier. But Schubert *et al* though that the role of the carrier is activating some reactants, or interacting with the active species which can activate the active site of reactants. [19] Thus, a series of catalysts formed by supporting noble metals on transition metals or rare earth oxides, such as Au/CeO_2, [20] Au/Co_3O_4-CeO_2, [21] Ag/MnOx-CeO_2, [22] Pt/MnOx-CeO_2, [23] Pd/TiO_2, [24] Pt/TiO_2, [25], Pt/Fe_2O_3, [26] Au/ZrO_2, [27] and PdMn/Al_2O_3 [28] were developed. It was excited that the activity of these catalysts were obviously improved through changing supports and active species, and the temperature for complete conversion was largely reduced.

In general, noble metals (Pt, Pd, Ru, Rd and Ir) are typically used as the activity species of catalysts, because of their high activity and good stability. However, noble metals are expensive and vulnerable to poisoning, and the applications of these catalysts in commercial scale have been seriously impeded by these serious concerns. Besides, in some cases, Pt- or Pd-based catalysts exhibit a poor activity below about 200°C (Table 2) and are unable to clean up gaseous pollutants at or close to ambient temperatures. In recent years, there is increasing interest in searching for alternatives to noble metals as the active catalyst components. Some metal oxides [29] and transition metal perovskites [30] were investigated as substitutes for the noble metals, but their activities for catalytic HCHO combustion were unsatisfactory. Au is usually regarded as a poor catalyst for oxidation reaction, a behavior that was related to its electron configuration. [31] Haruta and coworkers [32] found that highly dispersed particles of metallic Au on supports exhibited a remarkable activity for CO oxidation at low temperature. This finding has initiated significant interest in Au catalysts. Much research has been concentrated on studying the catalytic properties of ultrafine Au nanoparticles and oxide-supported Au catalysts, which are proven to exhibit extraordinarily high activity in HCHO catalytic oxidation at moderate temperatures.

It is well known that the activity of Au catalysts mainly depend on the particle size of Au, loading content and support type. It is generally accepted that if the particle size of Au is less than 5 nm, Au catalysts will have high catalytic activity. Li *et al* [33] prepared a series of Au/Fe-O catalysts for catalytic oxidation of HCHO by co-precipitation method. The complete conversion of HCHO could be achieved around 80°C over such catalysts. Shen *et al* [20b] prepared a series of Au/CeO_2 catalysts by co-precipitation and subsequently calcined at different temperatures. The activity of the catalysts calcined at 300°C exhibited superior activity for catalytic oxidation of HCHO [Figure 4 (Left)]. However, when the catalysts were calcined at 400°C, their activity for HCHO catalytic oxidation decreased [Figure 4 (Right)]. TEM characterization showed that the size of Au activity [Figure 5 (Left)]. When calcining the catalysts at 700°C, the large Au nanoparticles (≥50 nanoparticles is about 10-20 nm, which may mainly account for the decrease of the catalytic nm) appeared [Figure 5 (Right)], resulting in the further decrease of the catalytic activity.

Table 2. Combustion of HCHO [31]

Catalyst	T_{50}/(°C)	T_{90}/(°C)	Catalyst	T_{50}/(°C)	T_{90}/(°C)
Ru/CeO_2	< 150	150	Ru/ZrO_2	188	276
Pd/CeO_2	< 150	181	Ru/Al_2O_3	198	239
Rh/CeO_2	< 150	204	Ru/zeolite	210	320
Pt/CeO_2	< 150	304	Ru/TiO_2	212	364
Ir/CeO_2	207	281	CeO_2	238	297

HCHO=900 ppm, SV=20000 h^{-1}.

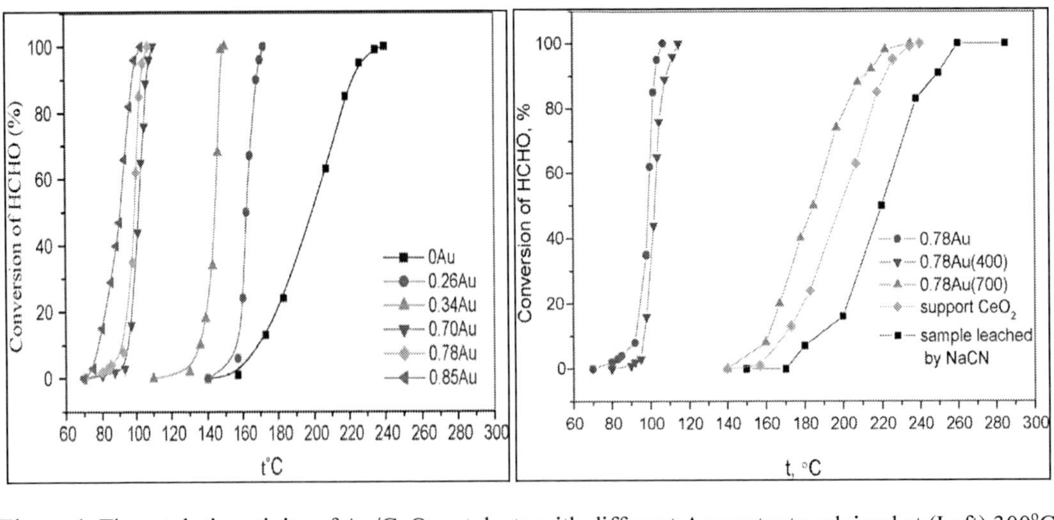

Figure 4. The catalytic activity of Au/CeO_2 catalysts with different Au contents calcined at (Left) 300°C and (Right) 400 °C and 700 °C for 2 h for HCHO catalytic oxidation [20b].

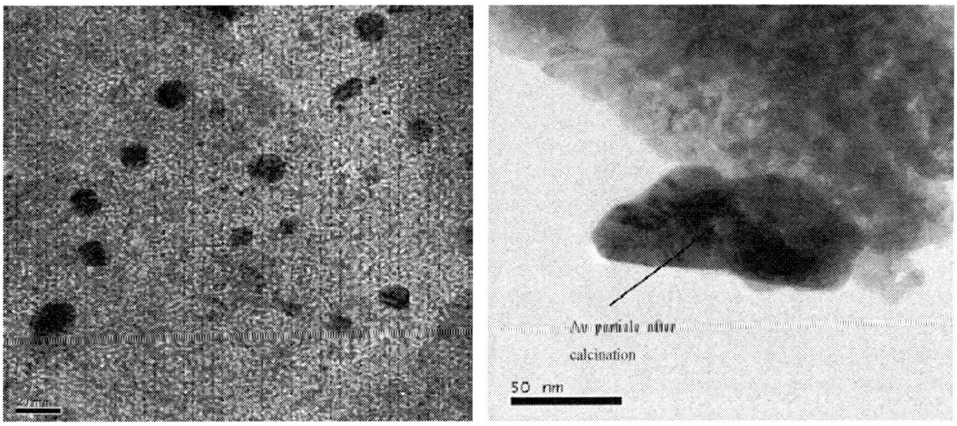

Figure 5. TEM image of Au/CeO_2 catalysts calcined at (Left) 400°C and (Right) 700°C. The dark spots are Au nanoparticles [20b].

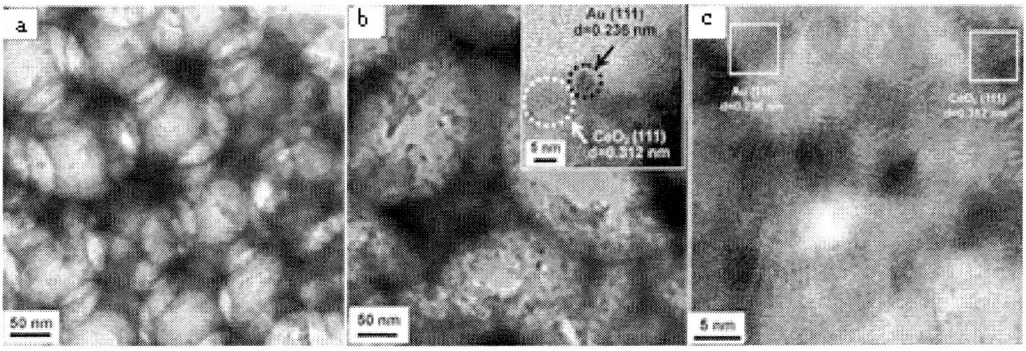

Figure 6. TEM (a) and HRTEM (b and c) images of 3DOM Au /CeO_2 synthesized via a gas bubbling-assisted deposition precipitation method. The black and white circles in the insert of (b) and the white rectangles in (c) clearly show the lattice fringes of Au (111) and CeO_2 (111), firmly suggesting the formation of small Au nanoparticles on 3DOM CeO_2 support [20a].

For the oxide supported Au catalysts, the aggregation of Au nanoparticles on powder catalyst supports may lead to the formation of large particles and reduce the active sites, which eventually lowering their catalytic activity. Zhang *et al* prepared three-dimensional ordered macroporous (3DOM) Au/CeO_2 catalysts [Figure 6(a)], [20a] which shows unique 3DOM structures with interconnected networks of spherical voids favoring less aggregation and good distribution of small Au nanoparticles [Figure 6(b), (c) and (e)], and their catalytic activity for HCHO catalytic oxidation was greatly improved with 100% HCHO conversion at temperature as low as 75°C, approximately 25°C lower than previously reported Au/CeO_2 catalysts without porous structure. In addition, unique porous geometry benefits to the loading of key catalyst species and prevents the aggregation of Au nanoparticles. XPS measurements showed that the Au valence state in 3DOM Au/CeO_2 catalysts before and after used for HCHO oxidation was quite different. In the freshly prepared 3DOM Au/CeO_2 catalyst, Au^0 and Au^{3+} coexist; while after being used for HCHO catalytic oxidation, the Au valence state changed from Au^{3+} to Au^0 (Figure 7), indicating that Au valence state is crucially important to the catalytic activity of 3DOM Au/CeO_2 catalyst.

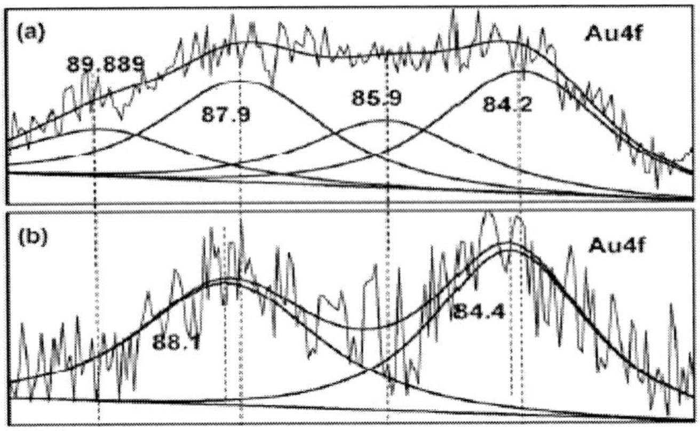

Figure 7. Au4f XPS spectra of Au ~1.0 wt%/3DOM CeO_2 catalysts with 80 nm pore sizes: (a) the as-prepared by drying at 60°C before used for HCHO oxidation, (b) after used for HCHO oxidation [20a].

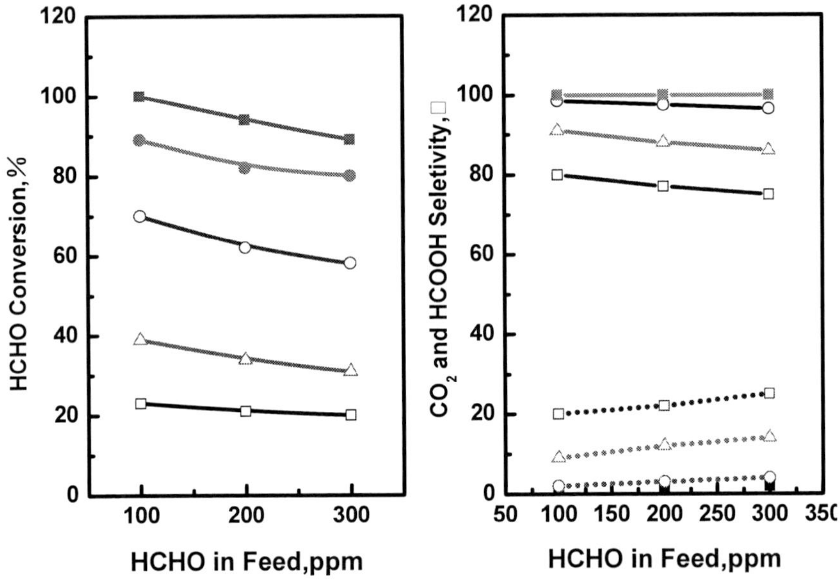

Figure 8. HCHO conversion at different temperatures and feed compositions (■) 150, (●)125, (○)103, (△)83, (□) 63°C(left). Selectivity to CO_2 and HCHO at different temperatures and feed composition (■) 150, (●)125, (○)103, (△)83, (□)63°C. (—) CO_2 selectivity, (- - -) HCHO selectivity [34].

4.2. Kinetics and Mechanism of Catalytic Oxidation of HCHO

The catalysts with high catalytic activity are greatly needed for removal of HCHO in polluted air. In order to clarify the factors that affect the catalytic activity of hydrophobic catalysts, Chuang *et al* [34] investigated the catalytic property of hydrophobic catalysts for HCHO catalytic oxidation. It was found that HCOOH is one of the intermediates during HCHO catalytic oxidation process. Figures 8 showed the catalytic activity and selectivity of hydrophobic catalysts for HCHO catalytic oxidation. When the concentration of HCHO is 100 ppm, the complete conversion for HCHO catalytic oxidation can be achieved at 150°C.

With increasing temperature, the selectivity to CO_2 increased and a lower concentration of HCOOH intermediate is also produced. When the reaction temperature is over 125°C, 100% selectivity to CO_2 is achieved.

Chuang and coworkers suggested that two processes including complete oxidation and partial oxidation were involved in HCHO catalytic oxidation. They proposed the mechanism of HCHO catalytic oxidation,[35] and a schematic expression of the reaction step was described by the following sequence according to the Mars-van Krevelen mechanism, which suggested that HCHO oxidation involved the processes of surface adsorption, desorption, and reaction for both reactants and products. The surface intermediate (HCOOH) could either be desorbed into the gas phase or be further oxidized into (CO_2) and (H_2O). Among it, the reaction between step 3 and step 6 was assumed to be the rate-controlling step. [34]

$$\textit{Step}\ 1:\ O_2 + (\ ) \xrightarrow{k_1} O_2$$
$$\textit{Step}\ 2:\ O_2 + (\ ) \rightarrow 2(O)$$
$$\textit{Step}\ 3:\ HCHO + (O) \xrightarrow{k_{2a}} (HCOOH)$$
$$\textit{Step}\ 4:\ (HCOOH) \xrightarrow{k_3} HCOOH + (\ )$$
$$\textit{Step}\ 5:\ (HCOOH) + (O) \xrightarrow{k_4} (CO_2) + (H_2O)$$
$$\textit{Step}\ 6:\ HCHO + (O) \xrightarrow{k_{2b}} (HCHO \cdots O) \rightarrow (CO_2) + (H_2O)$$
$$\textit{Step}\ 7:\ (H_2O) \leftrightarrow H_2O + (\ )$$
$$\textit{Step}\ 8:\ (CO_2) \rightarrow CO_2 + (\ )$$

Brackets represent an active site. The oxygen concentration is kept at a level much higher than the HCHO concentration (P_{O2}~/P_{HCHO}=0. 21/(100-300×10^{-6})=2100-700), which can be considered as constant and $K_{H2O}P_{H2O} \rightarrow 0$. The rate expression may be described the following equation at above-mentioned condition. [34]

$$-r_{HCHO} = \frac{k_1 k_2 P_{O_2} P_{HCHO}}{k_1 P_{O_2} + \nu(1 + K_{H_2O} P_{H_2O}) k_2 P_{HCHO}} \qquad (1)$$

$$-r_{HCHO} = \frac{k_1 k_2 P_{O_2} P_{HCHO}}{k_1 P_{O_2} + \nu k_2 P_{HCHO}} \qquad (2)$$

$$-\ln(1 - X) = -[({}^{\nu k_2}\!/\!{}_{k_1 P_{O_2}})]XP_{HCHO^0} + k_2 t \qquad (3)$$

Table 3. Parameters derived from the rate equation (2) and activation energy according to Mare-van Krevelen mechanism [34]

Temp/°C	vk_2/k_1	k_2t	Activation energy	kJ/mol	Explanation
63	298.19	0.309	E_1	22.23	Surface adsorption energy of oxygen
83	481.42	0.580	E_2	37.29	Energy of HCHO oxidation
103	596.84	1.386	E_3-E_4	-19.98	Energy difference between HCOOH surface Desorption and reaction

Further carefully analyzing the the parameters of equation (2), it was very interesting that the ratio of vk_2/k_1 rate constant increases with the increase of temperature, which suggests that the reaction rate (k_2) between HCHO and adsorbed oxygen is more sensitive to the change of temperature compared with oxygen adsorption rate (k_1) on the surface. In addition, the data of activation energy demonstrates that the oxygen adsorption energy is much lower than the surface reaction energy. The rate-limiting step is the reaction between HCHO and surface oxygen. Except for the temperature which affects the selectivity and activity of HCHO catalytic oxidation, other parameters such as water vapor, $K_{H2O}P_{H2O}$, P_{O2}/P_{HCHO} affect the selectivity and activity of HCHO catalytic oxidation as well.

Different catalysts may follow different kinetics model of HCHO catalytic oxidation. Yang *et al* investigated the kinetics of catalytic oxidation HCHO over Au/CeO_2 catalyst. [36] Different concentration of HCHO catalytic oxidation activity were tested between 303 K and 413 K. Because no other products except H2O and CO2 could be observed in $HCHO+O_2 \rightarrow CO_2+H_2O$, they proposed a power-law kinetic model $r = -kC_{HCHO}^{\alpha}C_{O_2}^{\beta}$, where k is the temperature dependent rate constant, α and β are the reaction orders relating to HCHO and O_2, respectively. In Arrhenius form $k = k_0 e^{-E_a/RT}$, k_0 is the frequency factor and E_0 is the activation energy. When the concentration of HCHO in the feed gas was much less than that of O_2 ($C_{HCHO} \ll CO_2$), the term $C_{O_2}^{\beta}$ in $r = -kC_{HCHO}^{\alpha}C_{O_2}^{\beta}$ can be regarded as a constant, therefore, it can be combined into the term of the frequency factor k_0, and the final expression of the kinetic model can be represented as $r = -k_0 e^{-E_a/RT} C_{HCHO}^{\alpha}$. The catalytic reaction was conducted in a continuous-flow fixed-bed reactor (Φ8 mm, 0.2 g cat.), where there is no temperature and concentration gradient in the radial direction and no temperature gradient in the axial direction. According to the inlet and outlet concentration of HCHO, the parameters of k_0, E_0, and α in $r = -k_0 e^{-E_a/RT} C_{HCHO}^{\alpha}$ can be estimated by the Levenberg-Marquardt method, and a power-law kinetic model may be described the following equation: [36]

$$r = -0.46 \times e^{-14612/RT} C_{HCHO}^{\alpha} \quad (303K \langle T \langle 363K)$$

$$r = -295.78 \times e^{-14612/RT} C_{HCHO}^{\alpha} \quad (363K \langle T \langle 413K)$$

It can be noticed that the frequency factor and the activation energy vary greatly in different temperature ranges, implying possible different reaction mechanisms at low temperatures and high temperatures.

Different active species may lead to different kinetics of HCHO catalytic oxidation. Hao and coworkers reported catalytic oxidation process of HCHO over mesoporous Au/Co_3O_4 and Au/Co_3O_4-CeO_2 catalysts. [21] The proposed that the {110}facets of Co_3O_4 composed mainly of Co^{3+} ion are the main active facets for HCHO catalytic oxidation, following the reaction steps listed below:

$$(1)\, Co^{3+} + HCHO \rightarrow Co^{2+} + {*CHO} + H^{+} \quad (2)\, {*CHO} + O^{\bullet-} \rightarrow HCOO*$$

$$(3)\, HCOO*^{-} + H^{+} \rightarrow HCOOH \quad (4)\, Co^{2+} + HCOOH \rightarrow Co^{2+} + HCOO* + H^{+}$$

$$(5)\, HCOO* + O^{\bullet-} \rightarrow HCO_3^{-} \quad (6)\, HCO_3^{-} + H^{+} \rightarrow H_2CO_3 \quad (7)\, H_2CO_3 \Leftrightarrow CO_2(g) + H_2O$$

The first step of HCHO catalytic oxidation is the adsorption of HCHO. Then, the HCOOH is produced by nucleophilic attack of surface active oxygen at the C-H of HCHO. Subsequently, the HCOOH is adsorbed on the Co_3O_4 {110} surface and form the •HCOO surface species. The second attack of surface active oxygen at the C-H of HCOOH to generate bicarbonate (HCO_3^-) species, and the bicarbonate (HCO_3^-) species combined with C-H to generate carbonic acid. Finally, the bound carbonic acid species decomposes into CO_2.

4.3. Different Active Species and Support with Different Catalytic Oxidation Mechanism

The kinetics and mechanism of HCHO interaction with solid surfaces is of particular significance for modeling catalytic process. If the active species are different, catalytic mechanism would be different. Liu *et al* investigated the catalytic mechanism of HCHO catalytic oxidation over 3DOM Au/CeO_2 catalyst. [37] The online GC characterization clearly monitored formation of HCOOH intermediate during the HCHO catalytic oxidation, which was further confirmed by FT-IR characterization. Moreover, the CO_2-TPD test for 3DOM Au/CeO_2 catalyst also showed that there were two peaks of CO_2 desorption and active Au species benefited to the desorption of CO_2. XPS spectra of Au4f of 3DOM Au/CeO_2 displayed the co-existence of two valence states of ionic Au^{3+} and metallic Au^0. Based on the above investigations, a catalytic mechanism of 3DOM Au/CeO_2 catalyst for HCHO catalytic oxidation is proposed.

Figure 9 illustrated the two catalytic mechanisms of HCHO catalytic oxidation in detail. Following the mechanism 1, when Au^{3+} is formed on the surface of Au/3DOM CeO_2 catalyst, the CeO_2 in contact with Au^{3+} may be partially reduced into Ce_2O_3. When HCHO is in contact with CeO_2 support, the absorption of HCHO onto the surface of CeO_2 support occurs. During this process, active oxygen may be transferred from Au_2O_3 to HCHO to form HCOOH and Au^0, and the HCOOH then changes into formate through the interaction with CeO_2 support, accompanying the losing of -H to form H_2O with free -OH absorbed on the surface of CeO_2 support. With the complete oxidation, HCOOH may be further converted into CO_2 and H_2O by continuously transferring the active oxygen to HCOOH. Meanwhile, the incomplete oxidation of HCOOH also occurs, this enables HCOOH to convert into carbonate and hydrocarbonate by losing one or two -H. When the lost -H interacts with the free -OH absorbed on the surface of CeO_2 support, H_2O is generated.

Following the mechanism 2, when Au exists in metallic state, there may be another process for HCHO oxidation catalyzed by metallic state Au^0, but this process is believed not the dominant process for the efficient conversion of HCHO into CO_2 and H_2O. When the similar absorption of HCHO onto CeO_2 support is happened, the active oxygen is transferred to HCHO to form HCOOH, and then HCOOH is converted into CO_2 and H_2O following a complete oxidation process. However, the incomplete oxidation also happens simultaneously to form carbonate and hydrocarbonate, which may deactivate the catalytic activity of Au/3DOM CeO_2 catalyst.

In general, 5A molecular sieves is not active for the oxidation of HCHO below 280°C. Sometime, silanols of zeolite could be the active sites in HCHO catalytic oxidation, which were related to the production of HCOOH (Figure 10). When 5A molecular sieves were ion-exchanged with Fe^{3+}, Fe-SiO_2 has poor activity for HCHO catalytic oxidation. But Fe^{3+} ions could be embedded in the frameworks of zeolite 5A to form Lewis acid sites, which has strong capacity of absorbing negative ions and mainly contributes to the production of CO_2. The oxygen molecule and the carbonyl of HCHO and HCOOH molecule might be absorbed on one Lewis acid site simultaneously. Accordingly, CO_2 and H_2O were formed on Lewis acid site formed by Fe^{3+} ions on zeolite 5A (Figure 11). [38]

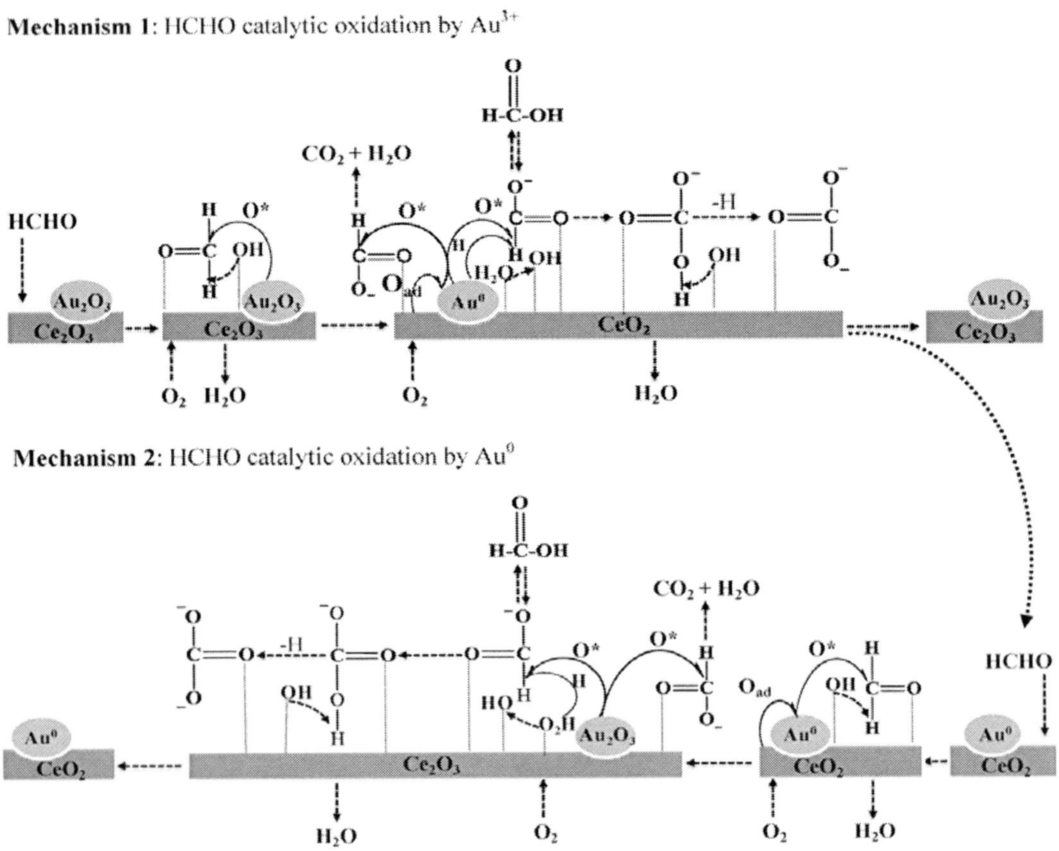

Figure 9. The proposed catalytic mechanism of 3DOM Au/CeO_2 catalyst for enhanced HCHO catalytic oxidation [37].

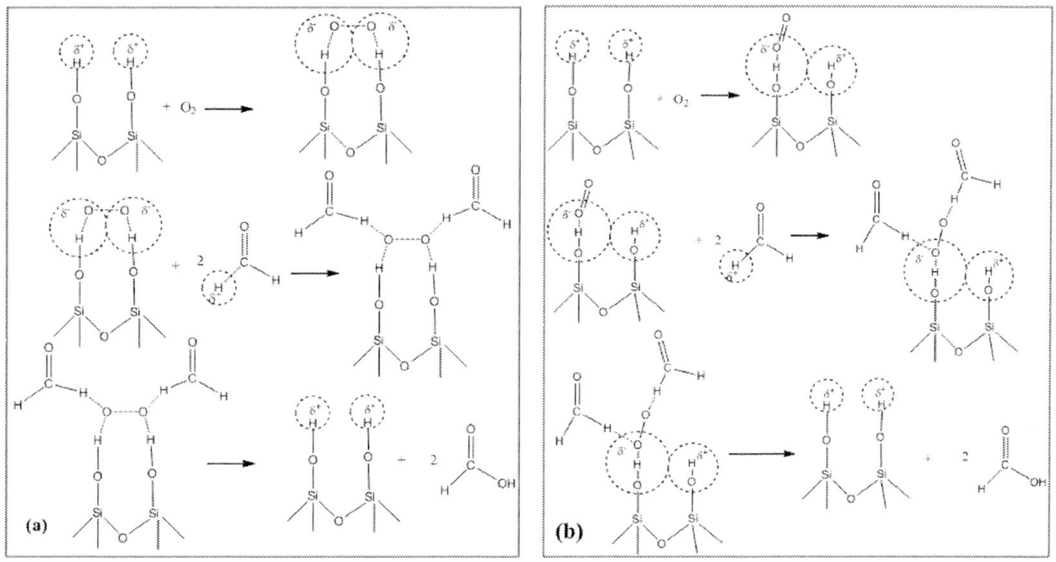

Figure 10. Mechanism of HCHO oxidation to HCOOH on silanols of zeolite 5A [38].

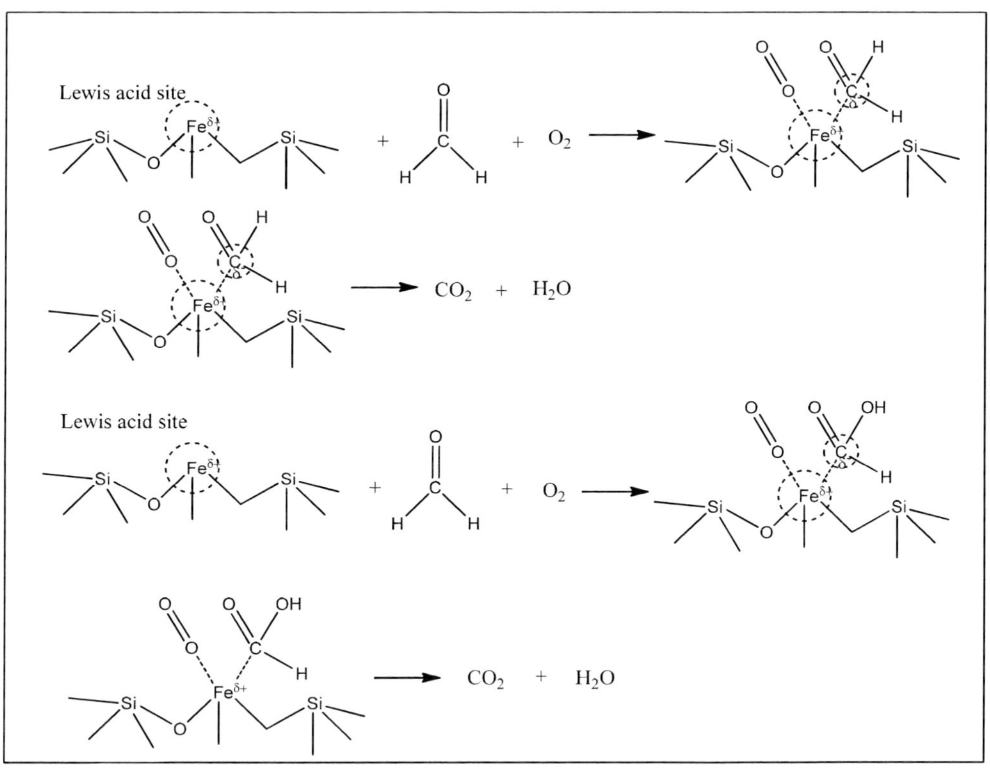

Figure 11. Mechanism of HCHO and HCOOH oxidation to CO_2 and H_2O on Lewis acid site formed by Fe^{3+} ions on zeolite 5A [38].

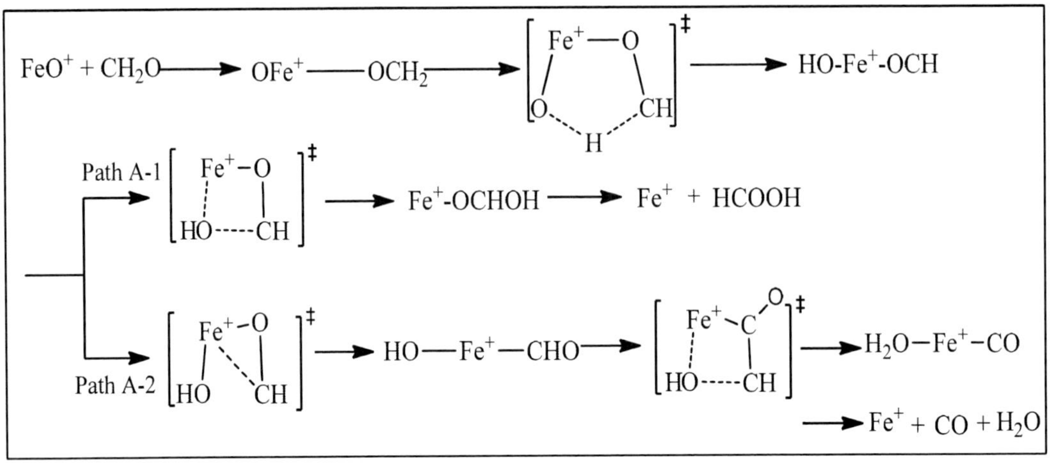

Figure 12. Mechanism of the HCHO to HCOOH and CO by FeO^+ complex [39].

Takashi and co-workers studied in detail the mechanism and energetics for the conversion of HCHO to HCOOH and HCOOH to CO_2 using FeO^+ complex as an oxidant from theoretical calculations at the B3LYP DFT (density functional theory) level. [39] HCHO oxidation mediated by FeO+ is expected to occur (Figure 12), in which FeO^+ and CHO_2 formed OFe^+-OCH_2. The reactant complex is converted to intermediate HO-Fe^+-OCH via a five-centered transition state for the H atom abstraction. After the formation of HO-Fe^+-OCH, there are two possible reaction pathways. One reaction pathways (Paths A-1) leads to HCOOH, the other pathways (Paths A-2) could lead to CO_2. Although the mechanisms of Paths A-1 and A-2 are quite different, according to DFT calculations, the CO complex in Path A-2 is 32.2 kcal/moL which is more stable than the HCOOH complex in Path A-1 in the quartet reaction pathway, which suggests that HCOOH can be released more easily than CO, and HCOOH should be further oxidized.

The conversion of HCOOH to CO follows the pathways of Paths B-1 and B-2 (Figure 13). In the first stages of Path B-1, the C-H bond of HCOOH is dissociated by the oxygen group of FeO^+ in a homolytic manner, which lead to the formation of intermediate of $HOFe^+$-OCOH•. Then, the O-H bond of the hydroxide ligand is dissociation, which results in the formation of CO and H_2O. In contrast, the O-H bond of HCOOH is cleaved by the oxygen group of FeO^+ in a heterolytic manner in the initial stages of Path B-2, and after that the final complex is formed via a five-centered transition state. In the reaction pathways for the conversion of HCOOH, the C-H bond dissociation by the oxygen group is the most difficult step, so is the C-H bond dissociation by hydroxide group. The C-H bond cleavage by the oxygen group in Path B-1 follows the reactant complex OFe^+-OCHOH; while that in Path B-2 follows intermediate HO-Fe^+-OCHO, as shown in Figure 12. The activation energy for the C-H bond cleavage in the sextet (quartet) reaction pathway along Path B-1 is -21.4 (-20.6) kcal/moL; while the C-H bond cleavage via the five-centered transition state along Path B-2 is -42.2 (-42.2) kcal/moL. But because the stabilization of HO-Fe^+-OCHO and OFe^+-OCHOH is different, these two reaction pathways are energetically favorable.

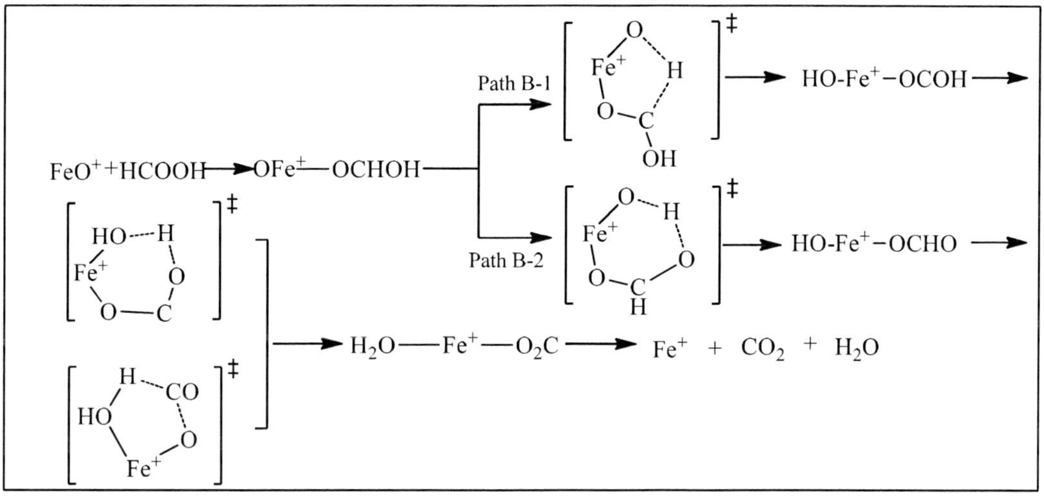

Figure 13. Mechanism of HCOOH oxidation to CO by FeO^+ complex [39].

CONCLUSION

With the broad applications of decorating and furnishing materials, HCHO is becoming one of the most concerned indoor air pollutants. On average, people spend as much as 87.2% of their time in indoor environment. When the concentration of HCHO exceeds the safe standard in indoor air, it may cause serious health problems including burning sensations in eyes and throat, nausea, difficulty in breathing, and even Leukemia or nasal cancer. It is very necessary to use highly active catalysts to remove HCHO in air at low temperatures.

All kinds of methods are used to eliminate HCHO except photocatalytic oxidation HCHO, such as room temperature adsorption, HCHO absorbers, negative ion and Low-temperature plasma technology, but they are hard t0 completely eliminate the HCHO due to its long emission time. Low temperature catalytic oxidation is an economical and effective way to eliminate indoor air contamination of HCHO. It implements the oxidative decomposition of HCHO under no heat source and light conditions, which also may save energy consumption and are widely used. On the other hand, there are still plenty of shortcomings in low temperature catalytic oxidation technology including the deactivation of catalyst, low catalytic efficiency, high conversion temperature and high cost. All of these reasons make the low temperature catalytic oxidation technology is difficultly industrialized. At present, both noble metal and non-noble metal catalysts are found to be very effective for HCHO catalytic oxidation. However, the ignition temperature of non-noble is higher than that of noble catalyst. In addition, water vapor and moist environment all make the non-noble metal catalyst loss activity. On the contrary, noble metal catalyst has high activity at room temperature, long life and the ability of resistance to water vapor and moist environment, which is a promising approach for eliminating HCHO.

Under the efforts of majority researchers, the active species, kinetic model, reaction mechanism, new type material system and structure of support already had been deeply research. It is reasonable to believe that there will find an efficient, economical and stable catalyst for elimination of HCHO in indoor air in the future.

REFERENCES

[1] Salthammer, T.; Mentese, S.; Marutzky, R., HCHO in the Indoor Environment. *Chemical Reviews* 2010, *110* (4), 2536-2572.

[2] Lide, D. R., *Handbook of Chemistry and Physics, 82ed.* CRC Press: Boca Raton, FL, 2001.

[3] Park, J. S.; Ikeda, K., Variations of HCHO and VOC Levels during 3 years in New and Older homes. *Indoor Air* 2006, *16* (2), 129-135.

[4] McGregor, D.; Bolt, H.; Cogliano, V.; Richter-Reichhelm, H. B., HCHO and glutaraldehyde and nasal cytotoxicity: Case study within the context of the 2006 IPCS human framework for the analysis of a cancer mode of action for humans. *Critical Review in Toxicology* 2006, *36* (10), 821-835.

[5] Kotzias, D.; Koistinen, K.; Kephalopulos, S.; Schlitt, C.; Carrer, P.; Maroni, M.; Jantunen, M.; Cochet, C.; Kirchner, S.; Lindvall, T.; McLaughlin, J.; Molhave, L.; de Oliveira Fernandez, E.; Seifert, B., The Index Project - Critical Appaisal of the Setting and Implementation of Indoor Exposure Limits in the EU. In *EUR-21590-EN*, EC-JRC, Ed. Ispra, 2005.

[6] (a) D.L., N.; M.M., C., *Lehninger Principles of Biochemistry*. Houndmills: Palgrave Macmillan, 2004; (b) Skrzydlewska, E., Toxicological and Metabolic Consequences of Methanol Poisoning. *Toxicology Mechanisms and Methods* 2003, *13* (4), 277-293.

[7] (a) Takeuchi, A.; Takigawa, T.; Abe, M.; Kawai, T.; Endo, Y.; Yasugi, T.; Endo, G.; Ogino, K., Determination of HCHO in Urine by Headspace Gas Chromatography. *Bulletin of Environmental Contamination and Toxicology* 2007, *79* (1), 1-4; (b) Canada, H., Proposed Residential Indoor Air Quality Guidelines for HCHO. Health Canada: Ottawa, 2005; Vol. H128-1/ 05-432E.

[8] Jones, A. P., Indoor Air Quality and Health. *Atmospheric Environment* 1999, *33* (28), 4535-4564.

[9] (a) Domingo-García, M.; Fernández-Morales, I.; López-Garzón, F. J.; Moreno-Castilla, C.; Pérez-Mendoza, M., On the Adsorption of HCHO at High Temperatures and Zero Surface Coverage. *Langmuir* 1999, *15* (9), 3226-3231; (b) Matsuo, Y.; Nishino, Y.; Fukutsuka, T.; Sugie, Y., Removal of HCHO from Gas Phase by Silylated Graphite Oxide Containing Amino Groups. *Carbon* 2008, *46* (8), 1162-1163.

[10] Fang, L. C.; Cai, P.; Li, P. X.; Wu, H. Y.; Liang, W.; Rong, X. M.; Chen, W. L.; Huang, Q. Y., Microcalorimetric and Potentiometric Titration Studies on the Adsorption of Copper by P. putida and B. thuringiensis and Their Composites with Minerals. *Journal of Hazardous Materials* 2010, *181* (1-3), 1031-1038.

[11] (a) Terigar, B. G.; Balasubramanian, S.; Boldor, D.; Xu, Z.; Lima, M.; Sabliov, C. M., Continuous Microwave-assisted Isoflavone Extraction System: Design and Performance Evaluation. *Bioresource Technology* 2010, *101* (7), 2466-2471; (b) Terelak, K.; Trybula, S.; Majchrzak, M.; Ott, M.; Hasse, H., Pilot Plant HCHO Distillation: Experiments and Modelling. *Chemical Engineering and Processing: Process Intensification* 2005, *44* (6), 671-676.

[12] (a) Akbarzadeh, R.; Umbarkar, S. B.; Sonawane, R. S.; Takle, S.; Dongare, M. K., Vanadia-titania Thin Films for Photocatalytic Degradation of HCHO in Sunlight. *Applied Catalysis A:General* 2010, *374* (1-2), 103-109; (b) Photong, S.;

Boonamnuayvitaya, V., Preparation and Characterization of Amine-functionalized SiO(2)/TiO(2) films for HCHO Degradation. *Applied Surface Science* 2009, *255* (23), 9311-9315; (c) Fu, P. F.; Zhang, P. Y.; Li, J., Photocatalytic Degradation of Low Concentration HCHO and Simultaneous Elimination of Ozone by-product Using Palladium Modified TiO(2) Films Under UV(254+185nm) Irradiation. *Applied Catalysis B:Environmental* 2011, *105* (1-2), 220-228; (d) Li, Y.; Jiang, Y.; Peng, S.; Jiang, F., Nitrogen-doped TiO_2 Modified with NH4F for Efficient Photocatalytic Degradation of HCHO under Blue Light-emitting Diodes. *Journal of Hazardous Material* 2010, *182* (1–3), 90-96.

[13] (a) Sekine, Y., Oxidative Decomposition of HCHO by Metal Oxides at Room Temperature. *Atmospheric Environment* 2002, *36* (35), 5543-5547; (b) Imamura, S.; Uchihori, D.; Utani, K.; Ito, T., Oxidative Decomposition of HCHO on Silver-cerium Composite Oxide Catalyst. *Catalysis Letter* 1994, *24* (3), 377-384; (c) Tang, X.; Li, Y.; Huang, X.; Xu, Y.; Zhu, H.; Wang, J.; Shen, W., MnO_x–CeO_2 Mixed Oxide Catalysts for Complete Oxidation of HCHO: Effect of Preparation Method and Calcination Temperature. *Applied Catalysis B: Environmental* 2006, *62* (3-4), 265-273.

[14] Liu, H.; Ye, X.; Lian, Z.; Wen, Y.; Shangguan, W., Experimental Study of Photocatalytic Oxidation of HCHO and Its by-products. *Research on Chemical Intermediates* 2006, *32* (1), 9-16.

[15] Wang, R.; Jiang, G.; Ding, Y.; Wang, Y.; Sun, X.; Wang, X.; Chen, W., Photocatalytic Activity of Heterostructures Based on TiO_2 and Halloysite Nanotubes. *Acs Applied Materials Interfaces* 2011, *3* (10), 4154-4158.

[16] Imamura, S.; Uematsu, Y.; Utani, K.; Ito, T., Combustion of HCHO on ruthenium/cerium(IV) oxide catalyst. *Industrial Engineering Chemistry Research* 1991, *30* (1), 18-21.

[17] Christoskova, S. G.; Danova, N.; Georgieva, M.; Argirov, O. K.; Mehandzhiev, D., Investigation of a Nickel Oxide System for Heterogeneous Oxidation of Organic Compounds. *Applied Catalysis A: General* 1995, *128* (2), 219-229.

[18] (a) McCabe, R. W.; Mitchell, P. J., Exhaust-catalyst Development for Methanol-fueled Vehicles: 1. A Comparative Study of Methanol Oxidation over Alumina-supported Catalysts Containing Group 9, 10, and 11 metals. *Applied Catalysis A: General* 1986, *27* (1), 83-98; (b) Lieske, H.; Lietz, G.; Spindler, H.; Völter, J., Reactions of Platinum in Oxygen- and Hydrogen-treated Pt/γ-Al_2O_3 Catalysts: I. Temperature-programmed Reduction, Adsorption, and Redispersion of Platinum. *Journal of Catalysis* 1983, *81* (1), 8-16; (c) Dorling, T. A.; Lynch, B. W. J.; Moss, R. L., The Structure and Activity of Supported Metal Catalysts: V. Variables in the Preparation of Platinum/silica Catalysts. *Journal of Catalysis* 1971, *20* (2), 190-201.

[19] Schubert, M. M.; Hackenberg, S.; van Veen, A. C.; Muhler, M.; Plzak, V.; Behm, R. J., CO Oxidation over Supported Gold Catalysts—"Inert" and "Active" Support Materials and Their Role for the Oxygen Supply during Reaction. *Journal of Catalysis* 2001, *197* (1), 113-122.

[20] (a) Zhang, J.; Jin, Y.; Li, C.; Shen, Y.; Han, L.; Hu, Z.; Di, X.; Liu, Z., Creation of Three-dimensionally Ordered Macroporous Au/CeO_2 Catalysts with Controlled Pore Sizes and Their Enhanced Catalytic Performance for HCHO Oxidation. *Applied Catalysis B: Environmental* 2009, *91* (1-2), 11-20; (b) Shen, Y.; Yang, X.; Wang, Y.; Zhang, Y.; Zhu, H.; Gao, L.; Jia, M., The States of Gold Species in CeO_2 Supported

Gold Catalyst for HCHO Oxidation. *Applied Catalysis B: Environmental* 2008, *79* (2), 142-148.

[21] Ma, C. Y.; Wang, D. H.; Xue, W. J.; Dou, B. J.; Wang, H. L.; Hao, Z. P., Investigation of HCHO Oxidation over Co_3O_4-CeO_2 and Au/Co_3O_4-CeO_2 Catalysts at Room Temperature: Effective Removal and Determination of Reaction Mechanism. *Environment Science Technology* 2011, *45* (8), 3628-3634.

[22] Tang, X.; Chen, J.; Li, Y.; Xu, Y.; Shen, W., Complete Oxidation of HCHO over Ag/MnOx–CeO_2 Catalysts. *Chemical Engineering Journal* 2006,118(1-2),119-125.

[23] Tang, X.; Chen, J.; Huang, X.; Xu, Y.; Shen, W., Pt/MnOx–CeO2 catalysts for the Complete Oxidation of HCHO at Ambient Temperature. *Applied Catalysis B: Environmental* 2008, *81* (1-2), 115-121.

[24] Huang, H.; Leung, D. Y. C., Complete Oxidation of HCHO at Room Temperature Using TiO2 Supported Metallic Pd Nanoparticles. *ACS Catalysis* 2011, *1* (4), 348-354.

[25] (a) Huang, H.; Leung, D. Y. C., Complete Elimination of Indoor HCHO over Supported Pt Catalysts with Extremely Low Pt Content at Ambient Temperature. *Journal of Catalysis* 2011; (b) Wang, L. F.; Sakurai, M.; Kameyama, H., Study of Catalytic Decomposition of HCHO on Pt/TiO_2 Alumite Catalyst at Ambient Temperature. *Journal of Hazardous Materials* 2009, *167* (1-3), 399-405.

[26] An, N. H.; Yu, Q. S.; Liu, G.; Li, S. Y.; Jia, M. J.; Zhang, W. X., Complete Oxidation of HCHO at Ambient Temperature over Supported Pt/Fe_2O_3 Catalysts Prepared by Colloid-deposition Method. *Jouranl of Hazardous Materials* 2011, *186* (2-3), 1392-1397.

[27] (a) Zhang, Y. B.; Shen, Y. N.; Yang, X. Z.; Sheng, S. S.; Wang, T.; Adebajo, M. F.; Zhu, H. Y., Gold Catalysts Supported on the Mesoporous Nanoparticles Composited of Zirconia and Silicate for Oxidation of HCHO. *Journal Molecular Catalysis A:-Chem ical* 2010, *316* (1-2), 100-105; (b) Hong, Y. C.; Sun, K. Q.; Han, K. H.; Liu, G.; Xu, B. Q., Comparison of Catalytic Combustion of Carbon Monoxide and HCHO over Au/ZrO_2 Catalysts. *Catalysis Today* 2010, *158* (3-4), 415-422.

[28] O'Shea, V. A. D. P.; Alvarez-Galvan, M. C.; Fierro, J. L. G.; Arias, P. L., Influence of Feed Composition on the Activity of Mn and PdMn/Al_2O_3 Catalysts for Combustion of HCHO/Methanol. *Applied Catalysis B:Environmental* 2005, *57* (3), 191-199.

[29] (a) Farneth, W. E.; Ohuchi, F.; Staley, R. H.; Chowdhry, U.; Sleight, A. W., Mechanism of Partial Oxidation of Methanol over Molybdenum(VI) Oxide as Studied by Temperature-programmed Desorption. *The Journal of Physical Chemistry* 1985, *89* (12), 2493-2497; (b) Lahousse, C.; Bernier, A.; Grange, P.; Delmon, B.; Papaefthimiou, P.; Ioannides, T.; Verykios, X., Evaluation of γ-MnO_2 as a VOC Removal Catalyst: Comparison with a Noble Metal Catalyst. *Journal of Catalysis* 1998, *178* (1), 214-225; (c) Papaefthimiou, P.; Ioannides, T.; Verykios, X. E., Combustion of Non-halogenated Volatile organic Compounds over Group VIII Metal Catalysts. *Applied Catalysis B: Environmental* 1997, *13* (3–4), 175-184.

[30] Spinicci, R.; Faticanti, M.; Marini, P.; De Rossi, S.; Porta, P., Catalytic Activity of $LaMnO_3$ and $LaCoO_3$ Perovskites towards VOCs Combustion. *Journal of Molecular Catalysis A: Chemical* 2003, *197* (1–2), 147-155.

[31] Bond, G. C., Gold: a Relatively New Catalyst. *Catalysis Today* 2002, *72* (1–2), 5-9.

[32] (a) Haruta, M.; Tsubota, S.; Kobayashi, T.; Kageyama, H.; Genet, M. J.; Delmon, B., Low-Temperature Oxidation of CO over Gold Supported on TiO_2, α-Fe_2O_3, and Co3O4. *Journal of Catalysis* 1993, *144* (1), 175-192; (b) Haruta, M.; Yamada, N.; Kobayashi, T.; Iijima, S., Gold Catalysts Prepared by Coprecipitation for Low-temperature Oxidation of Hydrogen and of Carbon Monoxide. *Journal of Catalysis* 1989, *115* (2), 301-309.

[33] Li, C.; Shen, Y.; Jia, M.; Sheng, S.; Adebajo, M. O.; Zhu, H., Catalytic Combustion of HCHO on gold/iron-oxide Catalysts. *Catalysis Communications* 2008, *9* (3), 355-361.

[34] Chuang, K. T.; Zhou, B.; Tong, S., Kinetics and Mechanism of Catalytic Oxidation of HCHO over Hydrophobic Catalysts. *Industrial Engineering Chemistry Research* 1994, *33* (7), 1680-1686.

[35] Chuang, K. T.; Cheng, S.; Tong, S., Removal and Destruction of Benzene, Toluene, and Xylene from Wastewater by Air Stripping and Catalytic Oxidation. *Industrial Engineering Chemistry Research* 1992, *31* (11), 2466-2472.

[36] Yang, X.; Shen, Y.; Bao, L.; Zhu, H.; Yuan, Z., Oxidation of Lean HCHO in Air over an Au/CeO_2 Catalyst and its Kinetics. *Reaction Kinetic Catalysis L* 2008, *93* (1), 19-25.

[37] Liu, B.; Li, C.; Zhang, Y.; Liu, Y.; Hu, W.; Wang, Q.; Han, L.; Zhang, J., Investigation of Catalytic Mechanism of HCHO Oxidation over Three-dimensionally Ordered Macroporous Au/CeO_2 Catalyst. *Applied Catalysis B: Environmental* 2012, *111–112* (0), 467-475.

[38] Xuzhuang, Y.; Yuenian, S.; zhangfu, Y.; Huaiyong, Z., Ferric ions doped 5A molecular sieves for the oxidation of HCHO with low concentration in the air at moderate temperatures. *Journal of Molecular Catalysis A: Chemical* 2005, *237* (1–2), 224-231.

[39] Yumura, T.; Amenomori, T.; Kagawa, Y.; Yoshizawa, K., Mechanism for the HCHO to Formic Acid and the Formic Acid to Carbon Dioxide Conversions Mediated by an Iron-Oxo Species. *The Journal of Physical Chemistry A* 2002, *106* (4), 621-630.

In: Formaldehyde
Editors: Chan Bao Cheng and Feng Hu Ln

ISBN 978-1-62257-214-4
©2012 Nova Science Publishers, Inc.

Chapter 5

INDOOR AIR MONITORING USING NEWLY DEVELOPED FORMALDEHYDE SENSOR ELEMENT AND PORTABLE MONITORING DEVICE

Yasuko Yamada Maruo*

NTT Energy and Environment Systems Laboratories, Kanagawa, Japan

ABSTRACT

I describe our developed formaldehyde sensor element, monitoring device, and its application to indoor air quality measurement. The sensor element we developed is made of a porous glass that is impregnated with both ammonium ions and 1-phenyl-1,3-butandione. The color of the sensor elements changes from colorless to yellow after exposure to formaldehyde with a peak wavelength of 415 nm. There is a linear relationship between the 415 nm absorbance of the sensor elements and the accumulated formaldehyde concentration. We estimated the formation reaction rate constant of lutidine derivatives (yellow dye) on the sensor element, and also estimated quantity of interference gases. We found that the reaction occurred sufficiently quickly for us to monitor hourly changes in the formaldehyde concentration. We also found that there were no interference gases under normal atmospheric conditions. The developed sensor element was a small, flat plate, pumping-free, and accumulated type, therefore we could install it for an arbitrary period in a space whose formaldehyde concentration we wished to determine. Then we could convert the absorbance change of the sensor element into the formaldehyde concentration using a preliminarily calculated calibration curve.

We also developed a portable device for formaldehyde monitoring, and carried out indoor air monitoring in several houses. The absorbance difference of the developed sensor element was measured at regular intervals in the monitoring device and converted into the formaldehyde concentration. This was possible because the lutidine derivative that was formed as a yellow product of the reaction between β-diketone and formaldehyde was stable in the sensor element. The detection limit was 5 ppb x hour. The monitoring device is small and easy to use and we used it to perform hourly formaldehyde monitoring using our monitoring device under several indoor conditions.

* E-mail: maruo.yasuko@lab.ntt.co.jp

We found that a high formaldehyde concentration could be measured in a room containing furniture and clothes. We also found that, although the formaldehyde concentration decreased rapidly when the room was ventilated, it recovered rapidly in several hours when we stopped the ventilation.

INTRODUCTION

Formaldehyde is a reactive gas emitted by various off-gassing sources, such as building materials, particle boards, adhesive and furniture. The World Health Organization (WHO) guideline for residential indoor air quality sets the formaldehyde exposure limit at an average of 80 ppb for 30 minutes [1]. Acute exposure to formaldehyde causes irritation of the eyes and upper airways, and long-term exposure to lower levels has been associated with an increased risk of developing respiratory illnesses [2]. Formaldehyde has been classified as a human carcinogen by the International Agency for Research on Cancer [3]. The outdoor emission sources of formaldehyde are mainly the gas emissions of cars and factories, however under atmospheric conditions, formaldehyde is readily photo-oxidized to carbon dioxide in sunlight, and the half-life of formaldehyde is several tens of minutes. Therefore the outdoor concentration is lower than the indoor concentration [4-9]. In contrast, the indoor emission sources of formaldehyde are resin products, adhesives, chipboard, fabric, tobacco smoke and heating devices [10-13]. It has been reported that the indoor formaldehyde concentration is usually 2-10 times higher than the outdoor concentration, and has a concentration range of several tens of ppb [14-19]. The major anthropogenic sources that affect humans are found in the indoor environment, and people spend 80 % of their time indoors.

A sensor that is both small and light is needed for estimating indoor levels of formaldehyde exposure. The DNPH method [20] is a well-known formaldehyde measurement technique that meets these requirements. However this analytical method requires many complex procedures including the storage of samples in a refrigerator, extraction using an organic solvent, and analysis using a chromatographic method. Multiple simultaneous measurements would not be easy to perform, and measurements may be impossible to undertake in a confined space, because the method requires a pumping unit. Nakano has reported a formaldehyde detection method called photoelectronic spectrophotometry, which employs both hydroxylamine hydrochloride and methyl-yellow impregnated paper [21]. Compton and Suzuki have reported a formaldehyde monitoring system based on the colorimetric reagent, 4-amino-4-phenylbut-3-en-2-one [22, 23]. Also, Kawamura has reported a formaldehyde monitoring reagent, 4-amino-3-hydrazino-5-mercaptotriazole [24]. Although formaldehyde can be detected using these four methods, they all require a device equipped with a pumping unit, and the measurement time is restricted to between 3 and 30 minutes. Therefore it is impossible to perform measurements in confined spaces, such as in drawers. In addition, it is impossible to measure the average formaldehyde concentration over a period of 1 day because the sampling time with these four methods is limited to 3 to 30 minutes as mentioned above.

Today a continuously operating air ventilation system is usually installed in individual family residences. Therefore, the formaldehyde concentration in newly built houses is less than that in older buildings, because of the improved quality of architectural materials and the installation of 24-hour air ventilation systems. However, large amounts of formaldehyde are emitted from various types of furniture and clothing [10-13]. Therefore it is assumed that although an empty newly built house has a low formaldehyde level, the formaldehyde concentrations in actual dwelling environments are high because of the presence of furniture or clothes.

The construction of energy-saving houses is a recent trend. The air exchange rate is sometimes reduced to improve indoor temperature control by using an air conditioner [25]. This may lead to an increase in the indoor formaldehyde concentration, which was formerly reduced by continuous ventilation [26]. Therefore, a portable formaldehyde monitor would be useful for determining a way of achieving both energy-saving and healthy (low formaldehyde level) conditions. Our purpose was to develop a simple and inexpensive portable instrument for the highly sensitive on-site monitoring of formaldehyde at ppb levels with a high temporal resolution. We found that when 1-phenyl-1,3-butandione was selected as the ß-diketone, the lutidine derivative that was a reaction product of ß-diketone and formaldehyde was stable in the porous glass for at least six months [27], whereas it was not stable in an aqueous solution [27], sol-gel glass [28], or on a cellulose sheet [29]. This enabled us to measure the accumulated formaldehyde concentration for arbitrary periods using the sensor element [30].

We also developed a formaldehyde monitoring device that incorporated the sensor element, small LED light sources, photodiodes and a slit to allow air diffusion [31]. The environmental formaldehyde enters the device through the slit, reaches the sensor element and reacts with the ß-diketone without the need for a pumping unit.

We were able to analyze the formaldehyde concentration continuously by measuring the change in the absorption of the lutidine derivative and converting the absorption difference into the formaldehyde concentration. We could also successfully monitor the formaldehyde concentration in several houses.

2. Sensor Element

We developed a sensor element for formaldehyde [27]. The sensor element was made of porous glass impregnated with both β-diketone and ammonium ions. We used three kinds of β-diketone; acetylacetone, 1-phenyl-1, 3-butanedione, and 1,3-diphenyl-1, 3-propanedione.

The three kinds of sensor element, which were initially colorless, turned yellow after exposure to formaldehyde, and absorption with a peak wavelength of 407-424 nm appeared. There were linear relationships between the 407-424 nm absorbance of the sensor element after exposure to formaldehyde, and the accumulated formaldehyde concentration.

The sensor element also worked cumulatively, and the absorbance changes of the three kinds of sensor elements were acetylacetone > 1-phenyl-1, 3-butanedione > 1,3-diphenyl-1, 3-propanedione, when exposed to the same concentration of formaldehyde in the atmosphere. The value of the absorbance change was about 1.0 for an accumulated exposure of 90-140 ppb × 24 hours for the acetylacetone or 1-phenyl-1,3-butandione. We also estimated the formation and decomposition reaction rate constants of lutidine derivative on the sensor element.

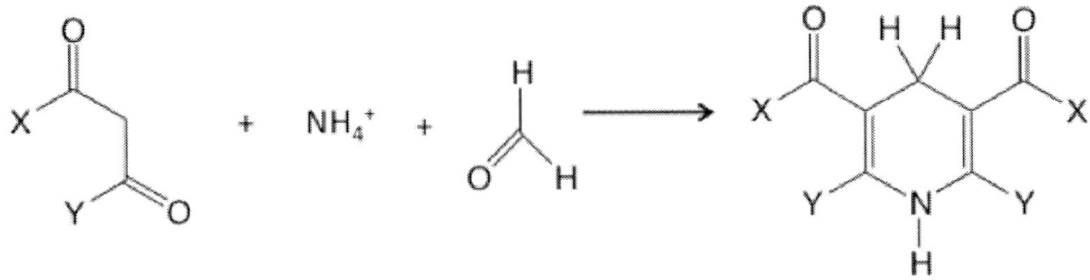

acetylacetone; $X=CH_3$, $Y=CH_3$
1-phenyl-1,3-butanedione; $X=C_6H_5$, $Y=CH_3$
1,3-diphenyl-1,3-propanedione; $X=C_6H_5$, $Y=C_6H_5$

Figure 1. Reaction formula of formaldehyde with β-diketone.

We found that the formation reaction rate constant on the acetylacetone sensor element was three times greater and the absorbance was 2.4 times greater than those of the 1-phenyl-1,3-butandione element. This means that the color of the acetylacetone element changes rapidly to a deeper yellow when it is exposed to formaldehyde. The decomposition reaction occurs on the acetylacetone sensor element, on the other hand, only the formation reaction occurs on the 1-phenyl-1,3-butanedione element, and the lutidine derivative is stable at least for three months. Therefore, we found that the 1-phenyl-1,3-butanedione and acetylacetone elements were suitable for long and short measurement periods, respectively.

Figure 1. shows the molecular formulas of acetylacetone, 1-phenyl-1, 3-butanedione, and 1,3-diphenyl-1, 3-propanedione, and the reaction scheme for β-diketone and formaldehyde.

2.1. Reaction between Formaldehyde and Three Kinds of β-Diketone

We examined the color-changing reaction of formaldehyde gas and β-diketone coexisting with both acetic acid and ammonium ions in the pores of the glass by exposing the samples to several different concentrations of formaldehyde gas for 24 hours. After exposure, the colorless sample became yellow. Figure 2. shows the change in the absorbance spectra of the acetylacetone element. Before exposure, there was a specific absorption at wavelengths below 340 nm, which was the absorption of the conjugated electrons of acetylacetone. After exposure, a specific absorption peak appeared at 407 nm, and it was clear that the absorbance increased as the formaldehyde exposure concentration increased. This peak is clearly separated from the absorption peak of acetylacetone. The acetylacetone coexisting with both acetic acid and ammonium ions in the pores was stable and there was no peak at 407 nm.

Therefore, the 407 nm peak that appeared would be a product of the reaction between acetylacetone and formaldehyde. If formaldehyde gas reacted directly with acetylacetone and ammonium ions in the glass pores then the lutidine derivative can be expected to form, which is reported to exhibit absorption at 400 nm in liquid [27], and the glass would turn yellow. And we realized a direct reaction between acetylacetone and formaldehyde on the pore's surface without using intermediate product of the acetylacetone method. Moreover, we could analyze 14 ppb of formaldehyde for 24hours exposure, without a gas-adsorption acceleration method (ex. pump).

Figure 3. shows the change in the absorbance spectra of the 1-phenyl-1, 3-butanedione element. Before exposure, there was a specific absorption at wavelengths below 370 nm, which was the absorption of the conjugated electrons of 1-phenyl-1, 3-butanedione.

After exposure, a specific absorption peak appeared at 414 nm and it was clear that the absorbance increased as the formaldehyde exposure concentration increased. The wavelength of this peak is separated from that of the absorption of 1-phenyl-1, 3-butanedione.

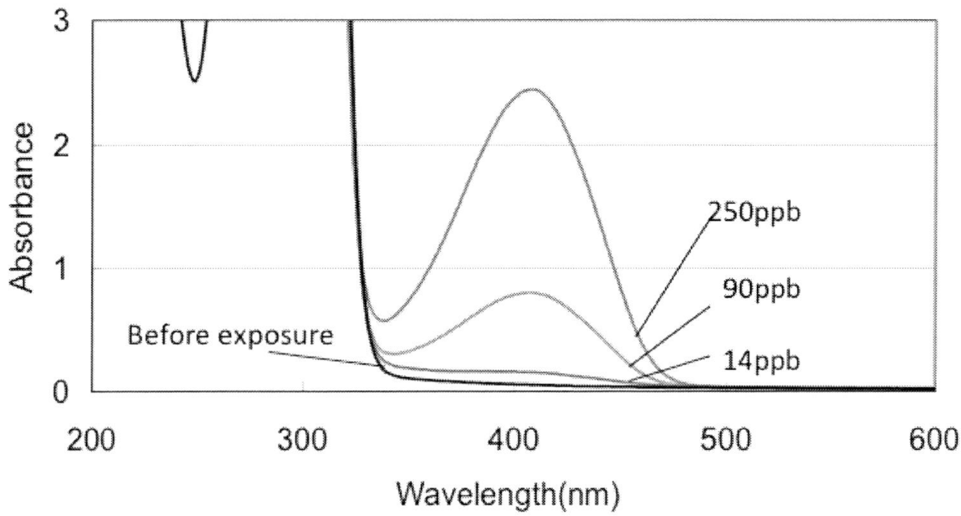

Figure 2. Change in absorption spectra caused by exposure to formaldehyde for acetylacetone impregnated sensor element (exposure time: 24 hours).

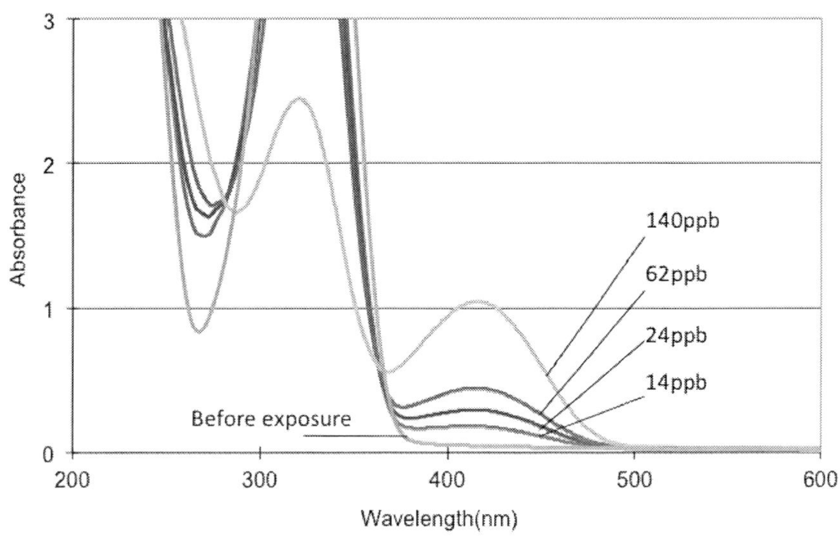

Figure 3. Change in absorption spectra caused by exposure to formaldehyde for a 1-phenyl-1,3-butandione impregnated sensor element (exposure time: 24 hours).

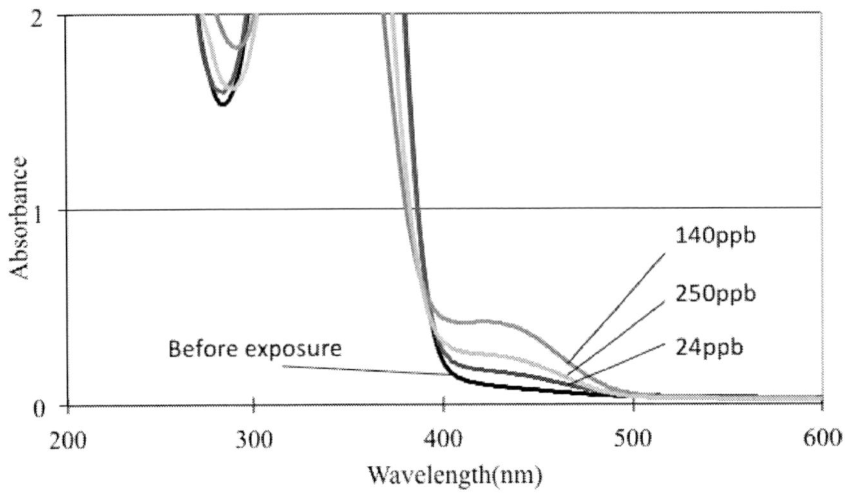

Figure 4. Change in absorption spectra caused by exposure to formaldehyde for a 1,3-biphenyl-1,3-propanedione impregnated sensor element (exposure time: 24 hours).

The 1-phenyl-1, 3-butanedione coexisting with both acetic acid and ammonium ions in the pores was stable, and there was no peak at 414 nm. As with acetylacetone, the formation of lutidine derivative (yellow dye) is expected. We realized a direct reaction between 1-phenyl-1, 3-butanedione and formaldehyde on the pore's surface without using the intermediate products of the acetylacetone method. The wavelength of the peak is 7 nm longer than that of acetylacetone, because of the substituent effect of the one benzene ring that is included in 1-phenyl-1, 3-butanedione. Figure 4. shows the change in the absorbance spectra of the 1,3-biphenyl-1,3-propanedione element. Before exposure, there is a specific absorption at wavelengths below 380 nm, which is the absorption of the conjugated electrons of 1,3-biphenyl-1,3-propanedione. After exposure, a specific absorption peak appeared at 425 nm and it was clear that the absorbance increased as the formaldehyde exposure concentration increased up to 140ppb. The wavelength of this peak is separated from that of the absorption of 1,3-biphenyl-1,3-propanedione. The 1,3-biphenyl-1, 3-propanedione coexisting with both acetic acid and ammonium ions in the pores was stable, and there was no peak at 425 nm. As with acetylacetone, the formation of lutidine derivative (yellow dye) is expected. We realized a direct reaction between 1,3-biphenyl-1,3-propanedione and formaldehyde on the pore's surface without using the intermediate products of the acetylacetone method. The wavelength of the peak is 18 nm longer than that of acetylacetone, because of the substituent effect of the two benzene rings that are included in 1,3-biphenyl-1,3-propanedione. For the absorbance decrease at 250ppb, we assumed that the lutidine molecules would change from one conformation with an absorption peak at 425 nm to another with an absorption peak around 400 nm [32]. For three kinds of β-diketone, we realized for the first time a direct reaction between β-diketone and formaldehyde on the surface of the glass pore, and specially the formation of the lutidine derivative (yellow dye) of 1,3-diphenyl-1, 3-propanedione have not reported anywhere yet. The formation of the lutidine derivative (yellow dye) of 1,3-diphenyl-1, 3-propanedione succeeds only on the pore surface. The mobility of molecules on the glass surface would be greater than that on paper, because there is a thin water layer on the glass surface. The presence of the water layer was confirmed by the existence of absorption peaks at 1490 and 1900 nm, which were caused by water.

2.2. Stability of Elements and Reaction Speed of Two Kinds of β-Diketone

We examined the stability and reaction speed of sensor elements containing β-diketone and ammonium ions and adsorbed formaldehyde gas. We used a pump unit to achieve the rapid adsorption of formaldehyde gas on the pore surface. After the formaldehyde gas adsorption had finished, the glass was stored in a polyethylene bag and its spectrum measured at the intended intervals. With the acetylacetone element, the glass gradually became yellow and then colorless. On the other hand, with the 1-phenyl-1, 3-butanedione element, the color was stable and the absorbance was saturated. Figure 5 shows the changes in absorbance of both the acetylacetone element at 407 nm and the 1-phenyl-1, 3-butanedione element at 414 nm. For the acetylacetone element, the absorbance rapidly increased for 2 hours and then decreased. For the 1-phenyl-1, 3-butanedione element, the absorbance rapidly increased for 2 hours and then saturated. We assumed a simple formation/decomposition reaction for the lutidine derivative (yellow dye) compounds as shown in equation (1).

$$2\beta\text{-diketone} + NH_4^+ + HCHO \xrightarrow{k1} L \xrightarrow{k2} D.C. \quad (1)$$

We fitted the peaks of the absorbance change plots of both the acetylacetone element and the 1-phenyl-1, 3-butanedione element to equation (1). The results are shown as a solid line in Figure 5. There are large quantities of β-diketone and ammonium ions in the porous glass. The product of k1, [β-diketone]^2 and [NH_4^+] is constant, and we replace the product with k'1. For the acetylacetone element, the obtained k1' (k1'=k1×[NH_4^+]×[β-diketone]^2) and k2 values are 3.16 h^{-1} and 0.0200 h^{-1}, respectively. For the 1-phenyl-1, 3-butanedione element, the obtained k1' and k2 values are 1.01 h^{-1} and 0.0000 h^{-1}, respectively.

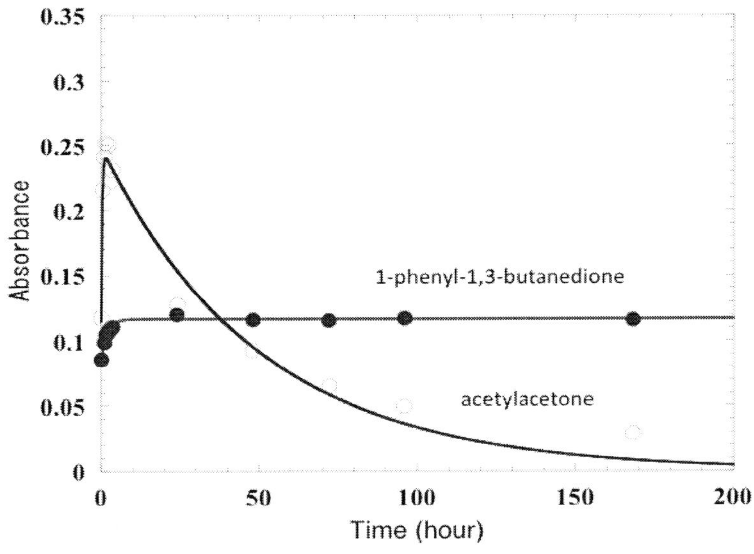

Figure 5. Relationship between standing time and peak absorbance of a β-diketone impregnated sensor element after formaldehyde gas adsorption.

Although the stability of the acetylacetone sensor element is not very stable, its sensitivity is higher than that of the 1-phenyl-1, 3-butanedione element, and the kinetic constant of the formation reaction is 158 times greater than that of the decomposition reaction. Therefore, we concluded that the acetylacetone sensor element is suitable for short-term measurements (e.g. less than 8 hours), and the 1-phenyl-1, 3-butanedione sensor element is suitable for long-term measurements (e.g. more than 8 hours).

3. Indoor Measurement

The developed sensor element was a small, flat plate, pumping-free and accumulated type, therefore we could install it in a space whose formaldehyde concentration we wanted to determine for an arbitrary period. Then the absorbance change of the sensor element could be converted into the formaldehyde concentration using the preliminarily calculated calibration curve.

We successfully measured the formaldehyde concentration in an enclosed space, because the developed sensor worked without a pumping unit. We also found that the formaldehyde concentration in a room containing furniture increased by 10% when the temperature increased by 1 °C.

3.1. Calibration Curve

The sensor elements were exposed to the gas by placing them in the Tedlar® bag at 20 °C and 60%RH. We balanced the formaldehyde gas in the Tedlar® bag with the formalin in the transpiration source by allowing 24 hours to elapse after the installation of the transpiration source. The transpiration rate was 1.68 mg/m^{2}h for our transpiration source, and our measurement showed that it took about 7 hours for the air in the Tedlar® bag to become balanced with the formalin in the transpiration source.

The exposure conditions for several formaldehyde concentrations involved using several diluted liquid formalin sources, and the concentration was checked with the 2,4-dinitrophenylhydrazine (DNPH)-derivatization method [33]. With the DNPH-derivatization method, formaldehyde was collected with 2,4-dinitrophenylhydrazone by drawing the air into the Tedlar® bag at 1.0 l/min through a Sep-Pak C cartridge (Waters, Millipore Corp, USA.). The 2,4-dinitrophenylhydrazone was extracted from the cartridges with 3 ml of HPLC grade acetonitrile, and the eluate was analyzed by high performance liquid chromatography (HPLC).

We used the above method to create a formaldehyde atmosphere with a concentration ranging from 18 to 335 μg/m^3. We exposed the 1-phenyl-1, 3-butanedione sensor element for 5 hours (6 samples) and 24 hours (3 samples) and then measured the absorbance changes at 414 nm. The exposure temperature was 24±1°C, and the humidity was 50±5%RH. A linear relationship was obtained between the accumulated formaldehyde concentrations (the product of formaldehyde concentration and exposure time) and absorbance changes for both exposure times as shown in Figure 6.

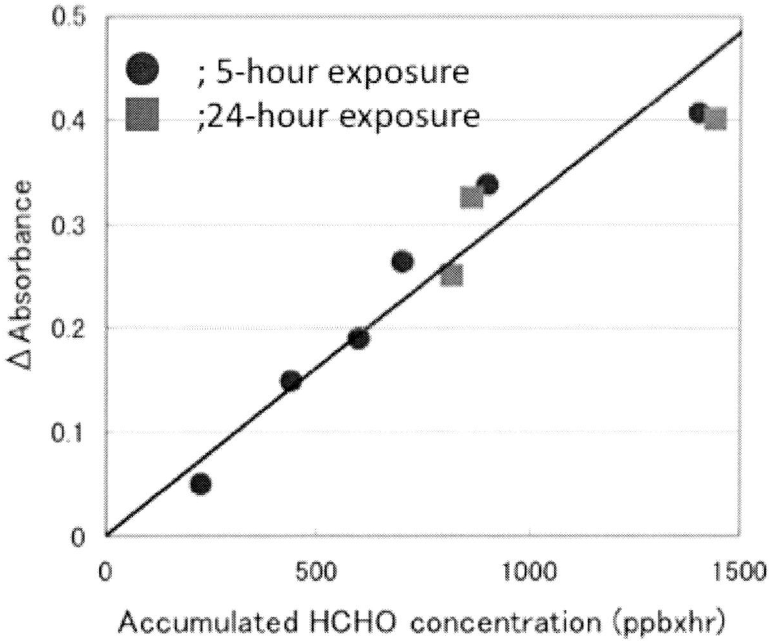

Figure 6. Relationship between the accumulated formaldehyde concentration and absorbance difference before and after exposure at 414 nm for the developed 1-phenyl-1,3-butandione sensor element.

The lines obtained by the least squares approximation method were the same for both exposure times indicating that the developed sensor element was an accumulated-type sensor, and the exposure time could be set at a desired length of for example, 6 or 8 hours. The calibration curve obtained from the relationship between the accumulated formaldehyde concentrations and absorbance changes is given by Eq. (2).

$$[HCHO] = (2.2\times10^{3} \times \Delta\ Abs.) / t \qquad (2)$$

In Eq.(2), [HCHO] indicates the formaldehyde concentration in ppb, Δ Abs. indicates the absorbance change at 414 nm, and t indicates the exposure time in hours. We were able to convert the absorbance change of a sensor element at 414 nm into a formaldehyde concentration using Eq. (2). We exposed the acetylacetone sensor element for 5 hours and then measured the absorbance changes at 407 nm.

The exposure temperature was 24±1°C, and the humidity was 50±5%RH. A linear relationship was obtained between the accumulated formaldehyde concentrations (the product of formaldehyde concentration and exposure time) and the absorbance changes. The calibration curve obtained from the relationship between the accumulated formaldehyde concentrations and absorbance changes is given by Eq. (3).

$$[HCHO] = (1.6\times10^{3} \times \Delta\ Abs.) / t \qquad (3)$$

In Eq.(3), [HCHO] indicates the formaldehyde concentration in ppb, Δ Abs. indicates the absorbance change at 407 nm, and t indicates the exposure time in hours. We were able to convert the absorbance change of a sensor element at 407 nm into the formaldehyde concentration using Eq. (3).

3.2. Residential House and Furniture Measurements

First we contacted volunteers who wanted to know the formaldehyde concentration in their houses or furniture, and asked them to measure the concentration using the 1-phenyl-1,3-butandione sensor element.

The samples consisted of 8 houses, 9 apartments, 12 pieces of furniture and 3 kinds of household item. An exposure time of 8 to 24 hours was arbitrarily selected. Measurements were carried out on an arbitrarily chosen day between September 30 and December 8 in 2007.

The temperature was in the 19±3 °C range, and the humidity was 70±20%RH. The air conditioning system was switched off at all locations. The formaldehyde concentration was measured in the living room and there were no smokers present. The experimental procedure was as follows.

(1) We measured the spectra of all the sensor elements, and distributed them to the volunteers.
(2) Each volunteer exposed the sensor element for an arbitrarily selected period of 8 to 24 hours at the location whose formaldehyde concentration they wished to know.
(3) After the period of exposure, each volunteer placed the exposed sensor element in an aluminum coated plastic bag to exclude light, and returned it to us.
(4) We measured the spectrum of each exposed sensor element, and calculated the formaldehyde concentration using the difference between the absorbance at 414 nm before and after exposure and the exposure time.

Table 1. shows the results for residential houses and apartments. All houses less than 5 years old had active ventilation systems, and the air change rate was 0.5 h^{-1}. On the other hand, none of the houses more than 6 years old had an active ventilation system. Houses less than 15 years old fall into the super-insulated house category, and houses that are more than 6 years old have no active ventilation system.

Therefore, the houses with the highest formaldehyde concentration are classified as super-insulated houses without an active ventilation system. In Japan, the Building Standards Act was revised in 2003, and stipulated that an active ventilation system had to be built into residential houses.

The Building Standards Act was also revised with regard to architectural materials and lumber (Japanese residential houses are mainly made of wood) was classified into four types with respect to formaldehyde transpiration rate. The recent trend is to use lumber with the lowest transpiration rate for residential houses, and the quality of lumber is improving year by year. Houses built less than 5 years ago are equipped with ventilation systems, and the quality of these systems is also improving every year.

Therefore, 5-year-old residential houses have lower formaldehyde concentrations than 10-year-old houses. However, active ventilation systems are unsuitable for energy-efficient houses, which will be the trend in the near future, because these systems consume a lot of energy. Therefore, in future it will become important to use active ventilation systems effectively while, for example, monitoring indoor air quality. The results for furniture are shown in Table 2.

Table 1. Formaldehyde concentration in various houses

measured location	sampling time	sampling date	Temperature (℃) (outside)	relative humidity (%) (outside)	HCHO ppb)	Features
residential house	8: 00~ 16: 00	9/29/2003	16.2	87	104	10years, without ventilation, super-insulated
residential house	8: 00~ 16: 00	9/29/2003	16.2	87	118	10years, without ventilation, super-insulated
residential house	8: 00~ 18: 00	10/9/2003	20.5	55	13	30years
residential house	0: 00~ 8: 00	10/9/2003	17.2	68	30	5years, with ventilation, super-insulated
residential house	0: 00~ 8: 00	10/11/2003	19.3	76	36	5years, with ventilation, super-insulated
residential house	9:00~ 17: 00	10/13/2003	18.6	61	24	5years, with ventilation, super-insulated
residential house	9:00~ 17: 00	10/13/2003	18.6	61	26	5years, with ventilation, super-insulated
residential house	22:00~ 7: 00	2007/10/18-19	10.8	83	29	5years, with ventilation, super-insulated
apartment	7: 00~ 15: 00	10/9/2003	20.4	57	46	10years
apartment	0: 00~ 8: 00	10/9/2003	17.2	68	37	20years
apartment	0: 00~ 8: 00	10/9/2003	17.2	68	12	20years, open-window
apartment	13:00~ 21:00	10/13/2003	17.7	67	24	10years
apartment	7: 00~ 16: 00	10/13/2003	18.5	60	12	10years, with ventilation
apartment	7: 00~ 16: 00	10/13/2003	18.5	60	15	10years, with ventilation
apartment	22:00~ 7: 00	2007/10/18-19	10.8	83	33	20years
apartment	7: 00~ 20: 00	10/17/2003	19.4	62	10	20years, open-window
apartment	9:00~ 18:00	10/27/2003	21.9	60	22	20years

As mentioned above, the developed sensor element operates without a pumping device; therefore we are able to measure the formaldehyde concentration in enclosed areas. There are many formaldehyde sources in drawers including chipboard, adhesive, and clothes, and a high concentration was assumed.

As shown in Table 2, the mean concentration was very high. However, the formaldehyde concentration outside the drawer was about one third of that inside. The mean formaldehyde concentration inside a chest of drawers was 32 ppb higher than that of desk drawers, and the main difference between a chest of drawers and desk drawers is that the former is used to store clothes, and clothes are relatively high transpiration sources as regards formaldehyde.

Table 2. Formaldehyde concentration in various pieces of furniture

measured location	sampling time	sampling date	temperature(℃) (outside)	relative humidity(%) (outside)	HCHO (ppb)	features
in a chest	8:00~16:00	2007/9/30	16.2	87	206	chipboard
in a chest	7:00~15:00	2007/10/10	20.4	57	65	chipboard
in a chest	8:00~18:00	2007/10/10	20.5	55	106	chipboard
in a chest	0:00~9:00	2007/10/11	16.4	69	104	chipboard
in a chest	0:00~8:00	2007/10/12	19.3	76	142	chipboard
in a chest	13:00~21:00	2007/10/14	17.7	67	128	chipboard
in a chest	22:00~7:00	2007/10/18-19	10.8	83	96	chipboard
in a desk	0:00~8:00	2007/10/10	20.5	55	44	chipboard
in a desk	9:00~18:00	2007/10/28	21.9	60	133	chipboard
in a closet	0:00~8:00	2007/10/10	17.2	68	30	chipboard
bookrack	0:00~8:00	2007/10/10	17.2	68	54	chipboard
in a drawer	22:00~7:00	2007/10/18-19	10.8	83	65	chipboard
shirt	10:00~10:00	2007/11/15-16	20	50	121	permanent-press shirt
stuffed toy	10:00~10:00	2007/12/8-9	20	50	13	man-made fabric
bookcover	10:00~10:00	2007/12/8-9	20	50	1	leather goods

3.3. Temperature Effect on Formaldehyde Concentration

To measure the temperature dependence of the formaldehyde concentration, we selected one room in one house with a relatively high formaldehyde concentration, and performed experiments from March to August 2008 with a humidity range of 60±10%RH. There is a linear relationship between the formaldehyde concentration and the temperature, and the slope was 0.042. From these results, we calculated that the increase in the transpiration was 10% for a temperature increase of 1 °C.

4. Indoor Air Monitoring Using Developed Portable Formaldehyde Monitoring Device

We have developed a portable device for formaldehyde monitoring, and have carried out indoor air monitoring in several houses. The absorbance difference of the developed sensor element was measured at regular intervals in the monitoring device and converted into the formaldehyde concentration. This was possible because the lutidine derivative that was formed as a yellow product of the reaction between ß-diketone and formaldehyde was stable in the sensor element. The device contained an LED as a light source and photodiodes as photo-detectors. It was sufficiently small (10 cm x 10 cm x 4 cm) to be installed at a desired location in the house. In addition, the device was able to monitor an enclosed area without a convection flow, because it did not use a pump for air sampling. The detection limit was 5 ppb x hour. The developed sensor device was small and easy to use and we successfully carried out hourly formaldehyde monitoring using our monitoring device under several indoor conditions. We found that a high formaldehyde concentration could be measured in a room containing furniture and clothes. We also found that, although the formaldehyde concentration decreased rapidly when ventilation was provided, it recovered rapidly in several hours when we stopped ventilating the room.

4.1. Monitoring Device

Figure 7. shows photographs of both the device and the sensor element, and also a cross-sectional view of the device. Humidity and temperature sensors were installed in the device to monitor the formaldehyde concentration, temperature and humidity simultaneously. A 10-bit analog-to-digital converter was also installed in the device. The performance of the device depends strongly on the performance of the analog-to-digital converter. Each side of the monitoring device was equipped with a pair of slits to allow air ventilation. We used an LED emitting at a wavelength of 415 nm as a light source to measure the absorbance of the lutidine derivative. We used two photodiodes to detect light intensity; one for the light that passed through the sensor element and the other for the light that did not pass through the sensor element as a reference. This allowed us to measure the absorbance with high accuracy, because the light intensity of an LED is usually dependent on temperature.

4.2. Artificial Gas Exposure

We constructed an artificial gas exposure system to expose the monitoring device to an arbitrary concentration of formaldehyde gas. We placed a Tedlar® bag (volume: 1×10^{-2} m^3) containing the formaldehyde diffusion source in a thermostatic chamber at 25 °C, and dry air (relative humidity $\leq 2\%$, and formaldehyde concentration ≤ 5 $\mu g/m^3$) was introduced into the bag at a rate of 250 ml/min. The humid output air including the formaldehyde was introduced into a second Tedlar® bag that was prepared for exposure to formaldehyde. Also, dry air (relative humidity $\leq 2\%$, and formaldehyde concentration ≤ 5 $\mu g/m^3$) was introduced into a water-bubbling bottle at 250 ml/min, and the humid output air was introduced into a second Tedlar® bag to control the humidity.

We placed the monitoring device equipped with the sensor element in the second Tedlar® bag and measured the absorbance of the sensor element every 10 min. To confirm the formaldehyde concentration in the second Tedlar® bag, the concentration was occasionally checked using the 2,4-dinitrophenylhydrazine (DNPH)-derivatization method.

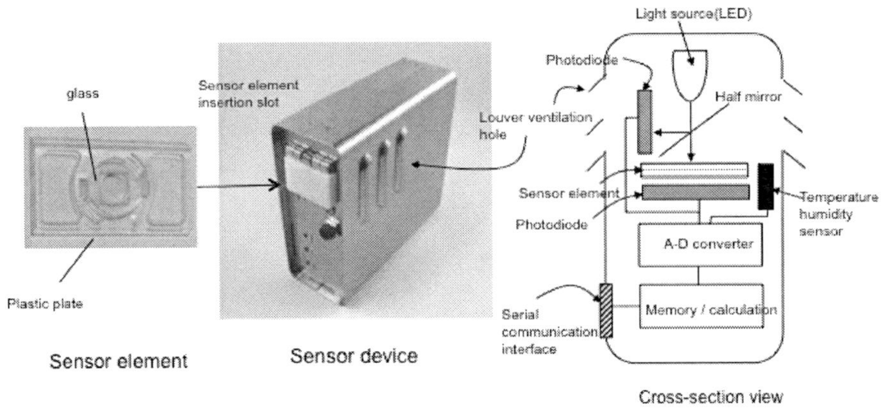

Figure 7. Photographs of the sensor element and monitoring device, and a cross-section of the monitoring device.

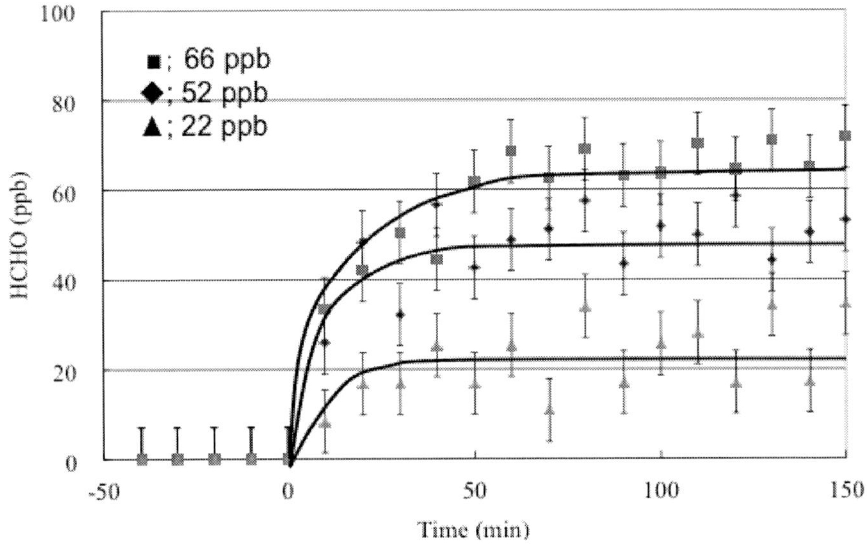

Figure 8. Changes in the output values of the formaldehyde concentration from the developed monitoring devices exposed to different formaldehyde concentrations (measurement interval: 10 min).

Atmospheres containing formaldehyde concentrations of 22, 52 and 66 ppb were used to measure the reaction time. The exposure temperature was 25 ± 1 °C and the humidity was 48 ± 2% RH. Figure 8. shows the output obtained every 10 min from each experiment when each device was set at three different formaldehyde concentration levels of 22, 52 and 66 ppb. The output values gradually increased, and a stable concentration of 94 to 100 % was reached after 1 hour of exposure. The output values are described as follows.

$$Dt = C \times (1 - \exp(-k1 \times t)) \quad (4)$$

Dt [ppb] indicates the calculated formaldehyde concentration from the absorbance measured by the developed monitoring device, C [ppb] is the exposed formaldehyde concentration, k1 [min^{-1}] is the reaction rate in the developed monitoring device, and t [min] is time. We used Eq. (4) to perform the least square fitting of the measured data as shown in Figures 8. The obtained k1 values were 0.049, 0.050 and 0.048 min^{-1} for formaldehyde concentration levels of 22, 52 and 66 ppb, respectively. Since the average k1 value is 0.049 min^{-1}, the calculated half-time is 14 min. Therefore, the amounts of formed lutidine derivative were calculated to be 76 % in 30 min, and 94 % in 1 hour.

4.2. Residential Apartment Measurement

We contacted volunteers who wanted to know the time series of formaldehyde concentration in their houses, and selected three apartments. The measurement period, location, floor area, temperature and humidity are listed in Table 3. The air-conditioning system was switched off in all three apartments. The formaldehyde concentration was measured in a closed room, and there were no smokers present. The experimental procedure was as follows:

(1) The sensor elements were packed into an aluminum-coated plastic bag to prevent gas exposure and light emission.
(2) The default values of time, date, measurement interval time (1 hour) were input into the device. If the device was once turned off, the values were stored in its memory and the measurement was automatically restarted when the device was turned on.
(3) Each volunteer set the device at a location whose formaldehyde concentration changes they wished to know, and placed the sensor element in the device and turned the device on. The measurement of the formaldehyde concentration started automatically and the resultant data were stored in the memory of the device.
(4) After 1 to 14 days measurement, each volunteer turned off the device, and the measurement data were extracted from the device to provide us with a time series of the formaldehyde concentration.

The monitoring results for the three apartments are shown in Figures 9 to 11 and a photograph of the monitoring device placed on a filing cabinet in a closed room in apartment3 is shown in Figure 12. For apartment1, the average concentrations were about 18 and 50 ppb in winter and summer, respectively. We have already reported that the increase in the transpiration of formaldehyde is 10% for a temperature increase of 1 °C. The obtained results support the previously reported value (10%$°C^{-1}$), because the average temperatures were 18 and 28 °C in winter and summer, respectively.

For apartment2, we opened the door of the closed room at the times indicated by arrow1 and arrow3 in Figure 10, and closed the door at the time indicated by arrow2 in Figure 10. The formaldehyde concentration decreased rapidly after we opened the door, and increased rapidly after it was closed.

In apartment3, we installed a new filing cabinet as shown in Fig. 12 at the time indicated by arrow1. The formaldehyde concentration increased by 10 ppb when the new furniture was installed.

Table 3. Specifications of measured apartment and measurement conditions

No.	location	floor	Area of house (m^2)	Area of room (m^2)	comment	period	Temp (C)	R.H. (%)
1	Isehara, Kanagawa	4/7	80	7.7	Large bookshelf	2008.12.28～ 2009.1.7 2009.9.2～ 9.3	18 28	46 60
2	Isehara, Kanagawa	1/2	60	7.7	1-year after renovation	2009.9.2～ 9.3	28	67
3	Atsugi, Kanagawa	4/5	80	7.7	Japanese room 2 years after renovation	2010.5.18～ 5.20	23	62

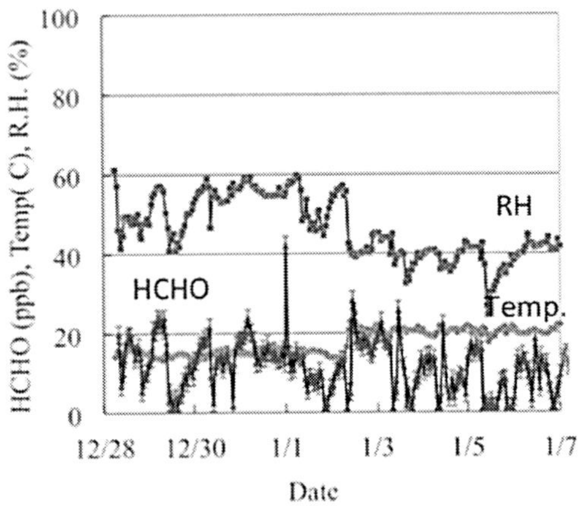

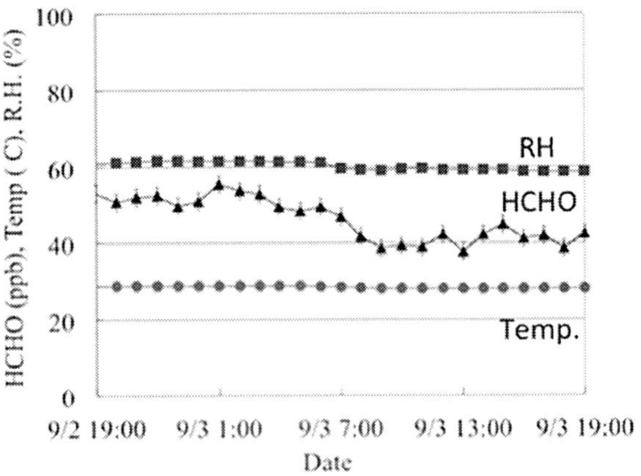

Figure 9. Time series variation of formaldehyde concentration, temperature and relative humidity in apartment1.

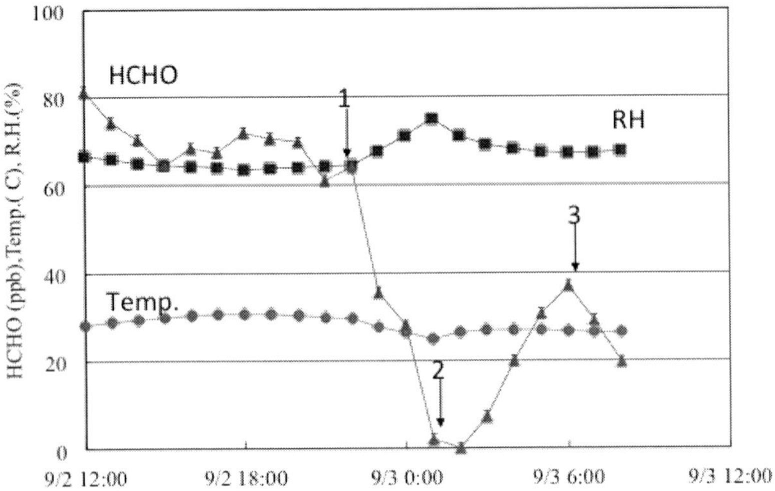

Figure 10. Time series variation of formaldehyde concentration, temperature and relative humidity in apartment2.

We opened the door of the room at the time indicated by arrow2 in Figure 11. The formaldehyde concentration decreased rapidly after we opened the door. For apartment1 and apartment3, although the temperature was almost constant, the humidity changed. We evaluated the influence of humidity on the formaldehyde concentration by using these data.

As shown in Figure 13, there is a linear relationship between the formaldehyde concentration and the relative humidity. However the slope is positive for apartment1 and negative for apartment3. The main difference is the style of the room; apartment1 is a Western style and apartment3 is Japanese style. The room in apartment1 has wooden floor and walls, whereas the room in apartment3 has a tatami (rush mat) floor and fusuma doors made of Japanese paper. However, the difference in the style of the room is just one factor. Since the influence of humidity on the formaldehyde concentration is very complex, we have to gather more data when considering the influence of humidity on the formaldehyde concentration.

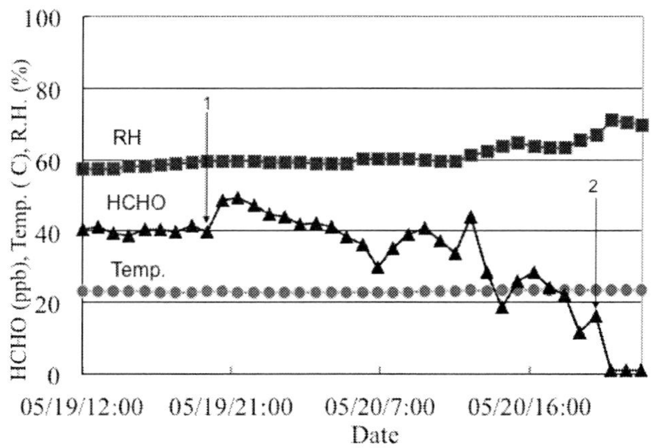

Figure 11. Time series variation of formaldehyde concentration, temperature and relative humidity in apartment3.

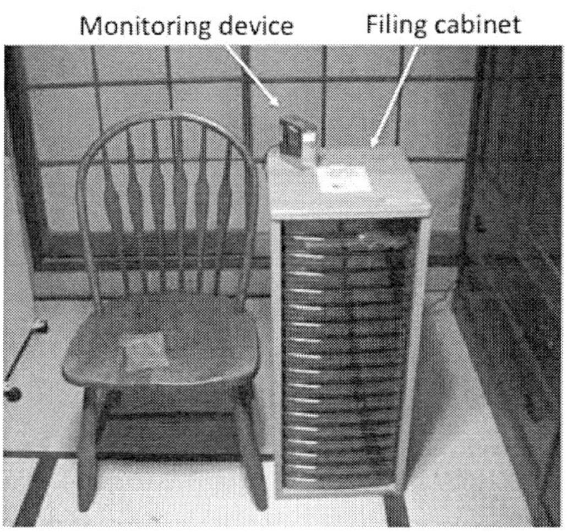

Figure 12. Photograph of developed monitoring device installed in a tatami room with furniture.

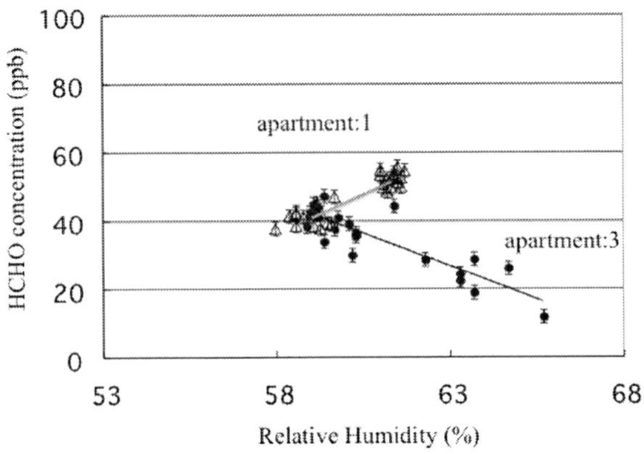

Figure 13. Relationship between formaldehyde concentration and relative humidity in apartment1 and apartment3 with a stable temperature.

Conclusion

I described the formaldehyde sensor element and monitoring device that we developed, and its application to indoor air quality measurement. The sensor element was made of a porous glass that was impregnated with both ammonium ions and ß-diketone. We examined three kinds of ß-diketone (acetylacetone, 1-phenyl-1,3-butandione, and 1,3-biphenyl-1,3-propanedione) and found that an acetylacetone sensor element is suitable for short-term measurements (e.g. less than 8 hours), and a 1-phenyl-1, 3-butanedione sensor element is suitable for long-term measurements (e.g. more than 8 hours). I also described indoor air measurements and the indoor air monitoring of hourly formaldehyde changes by using the developed sensor element and monitoring device, respectively. The developed sensor element operates without a pumping device; therefore we are able to measure the formaldehyde concentration in enclosed areas. We also estimated the effect of temperature or humidity on formaldehyde concentration, and found that the formaldehyde concentration in a room containing furniture increased by 10% when the temperature increased by 1 °C. As regards the effect of humidity, the results were very complex. There was a linear relationship between the formaldehyde concentration and the relative humidity, however the slope was positive for one apartment and negative for another apartment. We have to gather more data to consider the influence of humidity on the formaldehyde concentration.

References

[1] WHO-Air Quality Guidelines for Europe, WHO regional publications, ES No.23 (1987).

[2] Air Quality Guidelines-Second Edition, Chapter 5.8 Formaldehyde p. 1; WHO Regional Office for Europe, Copenhagen, Denmark (2001).

[3] IARC monographs on the evaluation of carcinogenic risks to humans, vol 88, formaldehyde, 2-butoxyethanol and 1-tert-butoxypropan-2-ol. WHO Press. Geneva (2006).

[4] N. Possanzini, V. D. Palo, A. Cecinato, *Atmospheric Environment*, 36, 3195 (2002).
[5] S. M. Correa, E. M. Martins, G. Arbill, *Atmospheric Environment*, 37, 23 (2003).
[6] O. Largiuni, R. Udisti, S. Becagli, R. Traversi, V. Maggi, E. Bolzacchini, P. Casati, C. Ugliett, S. Borghi, *Atmospheric Environment*, 37, 3849 (2003).
[7] M. Grutte, E. Flores, G. Andraca-Ayala, A. Baez, *Atmospheric Environment,* 39, 1027 (2005).
[8] M. Odabasi, R. Seyfioglu, *Atmospheric Environment*, 39, 5149 (2005).
[9] H. Tago, H. Kimura, K. Kozawa, K. Fujie, *Water, Air, and Soil Pollution*, 163, 269 (2005).
[10] F. Lipari, J. M. Dasch, W. F. Scruggs, *Environmental Science and Technology*, 18, 326 (1984).
[11] S. K. Brown, *Indoor Air*, 9, 209 (1999).
[12] T. J. Kelly, D. L. Smith, J. Satola, *Environmental Science and Technology*, 33, 81 (1999).
[13] S. Kim, J. A. Kim, J. Y. An, H. J. Kim, S. D. Kim, J. C. Park, *Indoor Air*, 17, 404 (2007).
[14] J. Zhang, Q. He, P. J. Lloy, *Environmental Science and Technology*, 28, 146 (1994).
[15] Baez, H. Padilla, R. Garcia, M. Torres, I. Rosas, R. Belmont, *Science of the Total Environment*, 302, 211 (2003).
[16] Hanoune, T. LeBris, L. Allou, C. Marchand, S. Le Calve, *Atmospheric Environment*, 40, 5768 (2006).
[17] Marchand, B. Bulliot, S. Le Calve, P. Mirabel, *Atmospheric Environment*, 40, 1336 (2006).
[18] Dassonville, C. Demattei, A. –M. Laurent, Y. Le Moullec, N. Seta, I. Momas, *Indoor Air,* 19, 314 (2009).
[19] H. Guo, N. H. Kwok, H. R. Cheng, S. C. Lee, W. T. Hung, Y. S. Li, *Indoor Air,* 19, 206 (2009).
[20] D. C. Lowe, U. Schmidt, D. H. Ehhalt, C. G. B. Frischkom, H. W. Numberg, *Environmental Science and Technology*, 15, 819 (1981).
[21] N. Nakano, M. Ishikawa, Y. Kobayashi, K. Nagashima, *Analytical Sciences*, 10, 641 (1994).
[22] B. J. Compton, W. C. Purdy, *Analitica Chimica Acta,* 119, 349 (1980).
[23] Y. Suzuki, N. Nakano, K. Suzuki, *Environmental Science and Technology*, 37, 5695 (2003).
[24] K. Kawamura, K. Kerman, M. Fujihara, N. Nagatani, T. Hashiba, E. Tamiya, *Sensors and Actuators*, B105, 495 (2005).
[25] Cosmos, http://www.osakaesco.jp/fair/2010/presen/cosmos2010.pdf.
[26] D. E. Hun, R. L. Corsi, M. T. Morandi, J. A.Siegel, *Indoor Air* 20, 196 (2010).
[27] Y. Y. Maruo, J. Nakamura, M. Uchiyama, *Talanta* 74, 1141 (2008).
[28] O. Bunkoed, F. Davis, P. Kanatharana, P. Thavarungkul, S. P. J. Higson, *Ana. Chim. Acta* 659, 251 (2010).
[29] Y. Y. Maruo, J. Nakamura, T. Miwa, Japanese Patent 2008-239209 (2008).
[30] Y. Y. Maruo, T. Yamada, J. Nakamura, K. Izumi, M. Uchiyama, *Indoor Air* 20, 486 (2010).
[31] Y. Y. Maruo, J. Nakamura, *Ana. Chim. Acta* 702, 247 (2011).

[32] Y. Y. Maruo, J. Nakamura, M. Uchiyama, *Journal of Environmental Chemistry*, 17, 413 (2007).
[33] R. W. Gillett, H. Kreibich, G. P. Ayers, *Environmental Science and Technology*, 34, 2051 (2000).
[34] JIS Methods for Determination of Formaldehyde in Fuel Gas, Nihon Kikaku Kyokai Press (2004) http://www.jisc.go.jp/app/pager?id=52887.

In: Formaldehyde
Editors: Chan Bao Cheng and Feng Hu Ln
ISBN 978-1-62257-214-4
©2012 Nova Science Publishers, Inc.

Chapter 6

UNUSUAL BEHAVIOR DURING THE ELECTROCHEMICAL OXIDATION OF FORMALDEHYDE

Mark Schell
Department of Chemistry, Southern Methodist University
Dallas, Texas, US

ABSTRACT

Formaldehyde is of great importance in industry and in research on polymers. Formaldehyde was and still is an intense research topic in electrochemistry. Consequently, it has played a key role in fundamental studies on the complex mechanisms for the electrochemical oxidation of small-oxygenated organic molecules. Following a survey of studies on the electrochemistry of formaldehyde, electrochemical behaviors not that well known are discussed. The oscillatory potential in response to the applied current was monitored during the electrochemical oxidation of formaldehyde. A sequence of temporal states was found consisting of intervals of periodic and chaotic behaviors. In part of the range the oxidation of formaldehyde exhibits a sequence of period doubling bifurcations. Anions usually inhibit chemical reactions. It is shown under potential control conditions and specified other conditions that nitrate, which typically inhibits reactions, can enhance the electrochemical oxidation of formaldehyde.

1. INTRODUCTION

Formaldehyde is an important commercial chemical with a major application in chemical synthesis. A less known but prevalent area of study is the electrochemistry of formaldehyde. Interest in how formaldehyde is oxidized at electrodes increased with the prospects of methanol fuel cells.[1] Originally it was thought that methanol might behave electrochemically in a manner similar to its behavior in solution chemistry where methanol is oxidized to formaldehyde, formaldehyde is oxidized to formic acid and formic acid is oxidized to carbon dioxide. [1,2] Separate studies were initiated on the electrochemical oxidation of methanol, formaldehyde, and formic acid.

In the remainder of this Introduction, we provide background to the study of the electrochemical oxidation of formaldehyde. We also provide some studies on another area of research for the electrochemical oxidation of formaldehyde, sensors. The number of studies referred to is admittedly incomplete. In the next section we provide an overview of electrochemical research on formaldehyde. It is meant to reveal how research produced many approaches undertaken to understand the electrochemical oxidation of formaldehyde and to increase its rate. In Sec. 3, nonlinear behavior exhibited by the electrochemical oxidation of formaldehyde is presented. This behavior includes period-doubling bifurcations, tangent bifurcations, and deterministic chaos. Finally, in Sec. 4, results of the effects of anions on the oxidation of formaldehyde are presented. Because surface intermediates are formed during the oxidation process, adsorbed anions decrease the rate of formaldehyde oxidation by occupying sites. Also studied is the case of a system containing formaldehyde and a large concentration of perchloric acid and a tiny amount of nitric acid. This tiny amount of nitric acid enhances the electrochemical oxidation of formaldehyde. The enhancement eventually seizes when the amount of nitric acid is increased, although it remains small.

It became apparent that the oxidation of methanol, formaldehyde, and formic acid at electrode surfaces took on much more complicated characteristics than the oxidation processes in solution chemistry. Although conditions can be found in which formaldehyde is a measurable intermediate of the electrochemical oxidation of methanol, [3] electrode surface intermediates dominate the process. Surface intermediates also form in the oxidation of formaldehyde and formic acid. Although in a general sense, there is an understanding of the mechanisms, a great deal of detail remains unknown.

One intermediate, surface bonded CO, [2] which bonds to the electrode surface in more than one way, occurs in the oxidation of all three organics. It has been well established that a reaction path exists in the oxidation of formic acid that does not include the CO intermediate. [2] Although less evidence exists, reaction paths without the CO intermediate are considered present in the mechanisms for the oxidation of formaldehyde and methanol. Reaction along paths without CO is faster at lower potentials. Surface CO is a long-lived intermediate at low potentials and severely inhibits the oxidation process by blocking the reaction paths that support faster reactions. Research has tried ways to reduce the inhibition by CO, prevent the formation of CO, or to cause surface CO to react at lower temperatures.

The desire to increase the rate of the electrochemical oxidation of formaldehyde and to understand the role of surface intermediates, as well as studies on the electrochemical oxidation of methanol and formic acid, opened up a fundamental area of research. Techniques were being developed for the study of a class of reactions, which may have been considered previously but if so left no footprint, that defined a new subfield in electrochemistry. No longer only considered as fuel-cell reactions, the electrochemical oxidations of formaldehyde, methanol, and formic acid are now studied as part of this subfield: complex electrochemical oxidation processes with more than one reaction pathway and several surface intermediates.

Formaldehyde has been studied less than methanol and formic acid in electrochemistry. In aqueous solution formaldehyde combines with water to form methylene glycol, but we will use the term formaldehyde. Formaldehyde is not stable in solution. One method of preparing formaldehyde solutions is to dissolve paraformaldehyde into the electrolyte at elevated temperatures. Using this method it is not clear whether dimers and trimers are formed especially without a stabilizers. Solutions are also made from formalin, an aqueous solution of formaldehyde and methanol. Methanol acts as a stabilizer. By using this method one can know what is in solution. However, this knowledge is not gained by reading the range of amounts on the bottle but rather through a separate analysis.

We now pay some attention to a somewhat large area in electrochemical research, which falls under the heading of sensors. Someone working with formaldehyde is almost certainly to eventually be subjected to eye and nasal irritation. Work on sensors are motivated by much more serious consequences of exposure to formaldehyde. Formaldehyde is classified as a human carcinogen that causes nasopharyngeal cancer. [4] Some studies have linked exposure to formaldehyde to an increase in leukemia [5].

One basic idea of designing a detector for formaldehyde is to begin with a simple electrochemical cell. Hämmerle did this and developed a detector for formaldehyde vapor. [6] It was created by incorporating the enzyme that catalyzes reaction of formaldehyde with NAD^+, formaldehyde dehydrogenase, into the working electrode. A linear response to formaldehyde was achieved up to 6 vppm. This work later led to the construction of an electrochemical cell within a plastic holding that contained the reference electrode, counter electrodes and the working electrode. [7] The working electrode consisted of disks of woven graphite in contact with a Pt wire on which a mediator was adsorbed. A buffer solution containing the enzyme surrounded the electrode; membranes were used to separate electrodes and prevent loss of enzyme. Formaldehyde vapor entered the housing and eventually the solution. Measurements showed a linear relation between current and formaldehyde up to 15 vvpm.

A sensor was made by immobilizing NAD^+- and recombinant formaldehyde dehydrogenase on the surface of a $Si/SiO_2/Si_3N_4$ structure. [8] The measurement in this case was capacitance. Calibration curves were measured for the capacitance vs the concentrations of formaldehyde. A linear relation was obtained for a concentration range from 10 μM to 20 μM.

An inorganic cell was constructed to serve as a sensor for the determination of formaldehyde. It was constructed by electrodepositing a nanostructured platinum palladium alloy in a Nafion film-coated glassy carbon electrode. [9] The electrode is the working electrode in a three-electrode cell. Voltammetric analyses revealed that the electrode displayed electrocatalytic activity for the oxidation of formaldehyde. The current response exhibited a linear relation with the formaldehyde concentration in the range of 10 μM to 1 mM. The detection limit was of 3 μM in acidic solution.

A Pd-modified TiO2 electrode provided characteristics that implied the electrode could be a good formaldehyde sensor. [10] The oxidation of formaldehyde produced peaks that occurred at potentials apparently lower than those obtained with other electrodes. Their values should be checked with respect to the standard Hydrogen electrode. A Ag/AgCl reference electrode was used in NaOH solutions. The latter reference electrode undergoes a large shift in alkaline solution.

The remainder of this paper is organized as follows. In Sec. 2, a range of studies is presented on the electrochemical oxidation of formaldehyde. A large portion of these studies focused on improving the current-potential characteristics of the oxidation process by examining the effects of different electrode surfaces. Studies that probed surfaced intermediates during the electrochemical oxidation of formaldehyde are also surveyed. In Sec. 3, a representative study is presented on nonlinear behavior in electrochemical processes:

potential oscillations observed in the electrochemical oxidation of formaldehyde under current control conditions. Several temporal behaviors are shown including mixed mode oscillations. Finally, typical and atypical effects of anions on the electrochemical oxidation of formaldehyde are discussed.

2. A Survey of Studies on the Electrochemical Oxidation of Formaldehyde

In this section we survey past research on the electrochemical oxidation of formaldehyde. Most research focused on finding ways to increase the current output at lower potentials, or on aspects of the mechanism including the intermediates. There was a period of time where there was disagreement on what was the surface species that inhibits the oxidation process. [2] It was probably in the early 1990s that it was agreed there was overwhelming evidence that it was CO. The best evidence came from experiments that employed some version of surface infrared. Much of the work was carried out using noble metal electrodes, mainly Pt. Much research examined whether better current-potential characteristics could be achieved by different electrode surfaces or modifications to a surface.

The electrochemical oxidation of formaldehyde was investigated at electrodes constructed of alloys. Alloys of Pd and group IB metals were studied in detail. [11] It was found that in alkaline solutions that the electrodes with greater Pd were highly active towards formate oxidation. The electrodes with substantial amounts of gold and copper were highly reactive to the oxidation of formaldehyde. The activity of alloys of Pt and silver and Pt and gold were studied. [12] A conclusion of the study was that the energy of activation is related to both the electronic and adsorption properties of the alloys.

Success at elevating the current in formaldehyde oxidation was achieved by using adatoms. [13,14] Atoms like Cu, Ag, Tl, Hg, Pb, As, Bi, Ge, and Sn were deposited on Pt electrodes. Under a wide range of conditions the current was elevated. Two mechanisms for enhancement apply. In one, the adatoms exert a geometric control. The geometry of their arrangement allows the oxidation of formaldehyde to proceed through a direct route but prevents the reaction pathway in which surface CO forms. A second mechanism occurs with adatoms like Ru which adsorbs oxygen containing molecules at a potential that is lower than the one in which this adsorption occurs on Pt. Since the oxygen material reacts with CO, reaction can proceed along the other paths.

Increases in the rate of the oxidation of formaldehyde in acid solution were achieved using catalysts consisting of Pt and Pt-Pd nanoparticles supported on single-walled carbon nanotubes. [15] The activity was greater than that achieved at other nanoparticle electrodes. The increases in activity were from the electrical properties of the single walled nanotubes as well as the high density of hydroxyl and carbonyl groups and other functional groups.

The electrochemical oxidation rates of formaldehyde, methanol and formic acid at carbon-supported platinum nanoparticle films were measured as a function of varying particle diameters in acid solution. [16] The rate of methanol oxidation decrease for nanoparticles with diameters less than 4 nm whereas the rate of frormic acid oxidation decreases. The rate of formaldehyde oxidation shows little sensitivity to the diameter of the nanoparticles. The formation of surface CO decreases with the size of the nanoparticle for all three oxidation processes. This decrease might arise from a geometric effect.

Silver nanostructures were produced in such a way that the morphology could be controlled. [17] The nanostructures produced included silver nanowires, silver nanorods, silver nanopolyhedra, and nanospheres. At glassy carbon electrodes modified by silver nanostructures, the rate of the oxidation of formaldehyde in 1.0 M KOH varied greatly with the morphology of the structure. There was significant activity at all nanostructures. The current peak obtained using nanorods was about 50 % greater than that using nanowires and both these structures produced currents significantly greater than the currents obtain using nanospheres and nanopolyhedra.

Spectroscopy played a key role in achieving an understanding about the presence and role of surface intermediates during the oxidation of methanol, formic acid and formaldehyde. Most of the earlier investigations helped to confirm that surface CO was the intermediate that reacted slowly at low potential and blocked the reaction paths associated with fast reaction rates. Use of potential-modulated reflection-adsorption infrared spectroscopy determined that surface CO has a consequential presence on Pt, Pd, and Rh electrodes during formaldehyde oxidation [18].

Using electromodulated infrared reflectance spectroscopy (EMIRS), surface CO was detected linearly bonded to the surface electrode atoms during the electrochemical oxidation of formaldehyde in a perchloric acid solution. [19] Bridge bonded and multibonded CO was also detected.

The electrochemical oxidation of formaldehyde at Pt(100) and Pt(110) electrodes in a perchloric acid medium was also investigated by (EMIRS). [20] Cyclic voltammetry was applied. Different oxidation processes were found to occur on each of the metals during potential scans in the forward direction. These processes involve either direct methylene-glycol oxidation or the formation of adsorbed CO species. The Pt(100) electrode is more active for the methylene glycol oxidation, which can be seen at low formaldehyde concentrations. Linear bonded CO was adsorbed on both electrodes, but multibonded and bridge bonded CO were only detected on the Pt(100) electrode. The surface CO blocks the oxidation of hydrated formaldehyde more on Pt (110) than on Pt(100).

Surface-enhanced infrared adsorption spectroscopy was employed to track surface species on a platinum electrode in a solution composed of 0.5 M sulfuric acid and 0.1 M formaldehyde during applications of cyclic voltammetry. [21] Linear and bridge CO as well as surface formate were detected. A conclusion of this work is that formate was an intermediate during the oxidation of formaldehyde.

3. Nonlinear Behavior during the Electrochemical Oxidation of Formaldehyde

The electrochemical oxidation of formaldehyde at electrode surfaces is a complex process. The presence of nonlinearities in the mechanism makes oscillatory behavior possible. Although studies that aim to detect intermediates in formaldehyde oxidation and studies that examine the effects of surfaces and electrode material on that process are far more numerous, there has been a steady stream of investigations of both current and potential oscillations. [22-

28] Waveforms representing different oscillatory states in the oxidation of formaldehyde are shown in Figure 1. These oscillations are potential oscillation obtained in current control experiments. Current oscillations can be obtained in potential controlled experiments but under the condition of a large solution resistance or applied resistance. The measurements are on a system consisting of a 500 ml, three-neck electrochemical cell held at 50^0 C. The cell contained 400 mls of a 0.4 M formaldehyde solution. (VWR. 37% formaldehyde, 5-10 % methanol. A quantitative analysis yielded 32.9% formaldehyde and 8.6% methanol (Hauser Laboratories, Boulder, CO).

The solution also contained 2.0 M H_2S0_4. The electrode was a disk with a diameter consisting of 5 mm Pt and 13 mm Teflon. The disk was rotated at 3500 rpm.

A number of different oscillatory states are represented in Figure 1. They are often referred to mixed mode oscillations because they consist of a mixture of large and small oscillation. Each member of the sequence of oscillations in Figure 1 can be represented by the notation 1^M, one large oscillation and M small oscillation. Such states are called principal states and sometimes the sequence is referred to a period adding sequence. The latter is a misnomer for highly technical reasons.

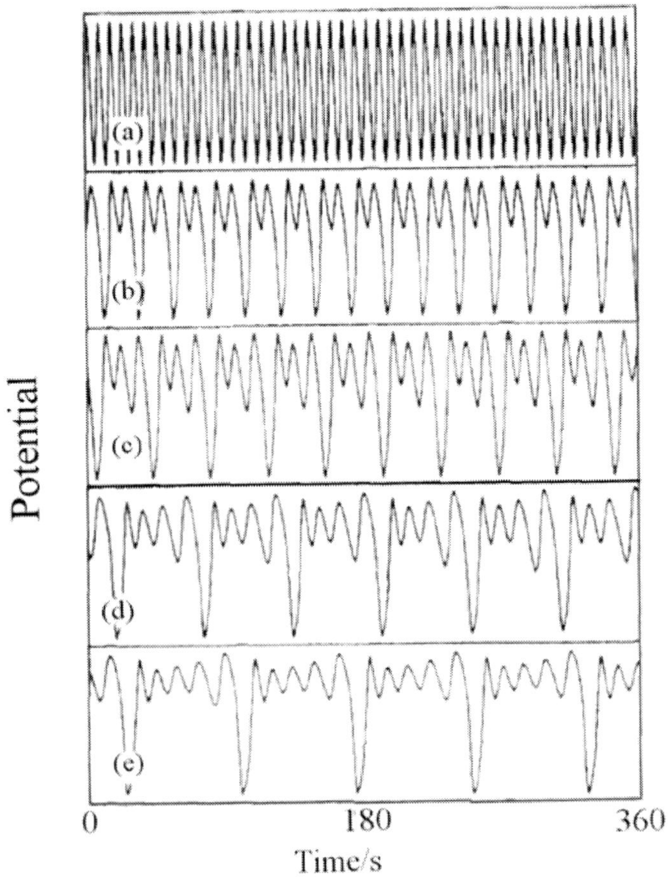

Figure 1. Measured potential oscillations. Potential range for each time series is 350 to 580 mV with respect to SCE. (a) 1^0 state, current (I) = 5.00 μA, (b)1^1 state, I = 3.5 μA, (c) 1^2 state I = 3.35 μA, (d) 1^3 state, I= 2.90 μA, (e) 1^4 state, I= 2.6 μA.

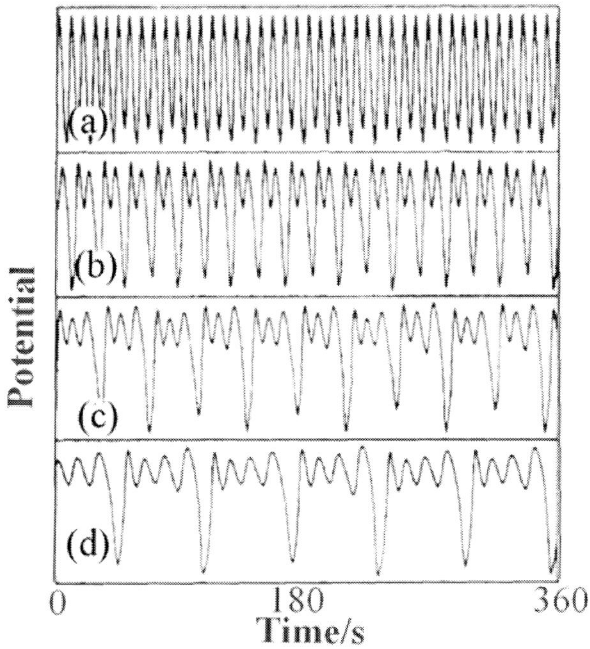

Figure 2. Potential is plotted against time. The waveforms are the subharmonics of the oscillatory states in Figure 1. (a) I = 3.80 μA, (b) state, I = 3.40 μA, (c) I = 3.10 μA, (d) I = 2.70 μA. Potential range same as Figure 1.

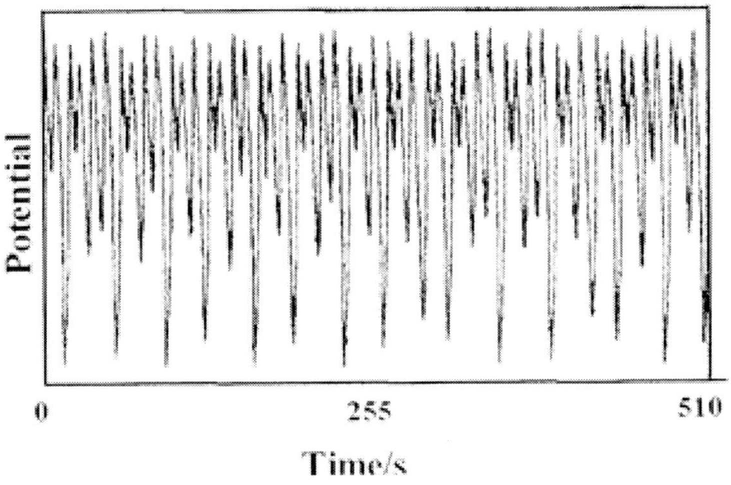

Figure 3. Chaotic oscillations. Potential same range as Figure 2. I = 3.38 μA.

Principal states are part of one of two possible sequences: A farey sequence contains principal states; members of the sequence can be added a certain way and when continued can, in principle, generate all the rational numbers. The principal states might also belong to a periodic-chaotic sequence in which periodic states are linked by period-doubling bifurcations. The oscillations in Figure 1 belong to the latter sequence.

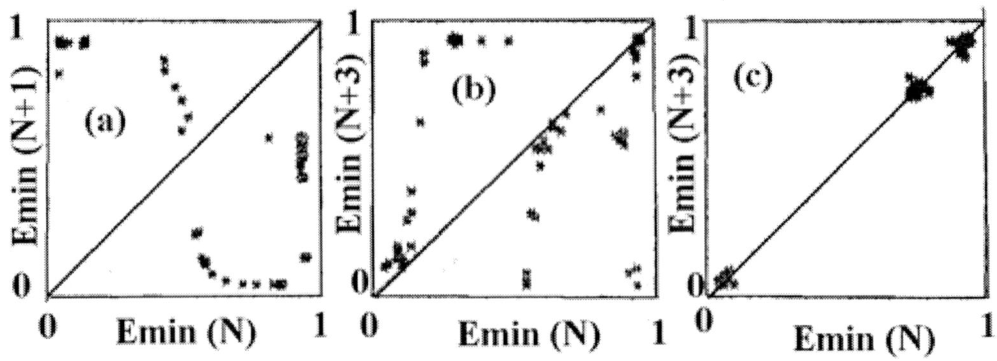

Figure 4. One-dimensional mappings. (a) Constructed from the time series in Figure 3. (a) The N+1 minimum in the time series is plotted against the Nth minimum. (b) and (c) the N + 3 minimum from time series is plotted against the Nth minimum.

When the current was decreased by small amounts from the values in Figure 1, each oscillatory state passed through a period doubling bifurcation. The 1^0 state, Figure 1(a), underwent a period doubling bifurcation which yields the oscillatory state in Figure 2(a). This state evolves with further decreases in current evolves to the one in Figure 1(b). To see that the states in Figure 2 are states with periods approximately twice as long as those in Figure 1, compare successive minima of the largest peaks in each Figure.

Decreasing the current from the value in Figure 1(b) causes the oscillatory state to undergo a period doubling bifurcation, which results in the oscillatory state in Figure 2(b).

The states in Figures 1(c) and (d) undergo period doubling bifurcations which give rise to the oscillatory states in Figures 2(c) and (d) respectively.

Chaos and other periodic states are expected to exist between the 1^1 and 1^2 states. A time series of chaotic potential oscillations is presented in Figure 3. Evidence that the time series is indeed chaos can be obtained by constructing a one-dimensional map. A map was constructed by plotting the (N+1)th potential minima vs the Nth potential minima from the time series. This one-dimensional map is in Figure 4(a). The iterates of the mapping outline a map function that performs stretching and folding which are characteristics of chaotic behavior.

Beginning at a current value that is greater than the values where the 1^2 exists, a bifurcation occurs from chaos to the 1^2 periodic state. This bifurcation is a tangent bifurcation. It can be seen through the construction of additional one-dimensional mappings. In Figure 4(b), a one-dimensional mapping is constructed by plotting the N+3 potential minimum in a time series vs the Nth potential minimum. The value of the current for the data in Figure 4(b) is less than the current used in Figure 4(a) and slightly greater than the current value where the 1^2 becomes stable. At this point the system is in a chaotic state. Note that the iterates outline a function in Figure 4(b) that is tangent to the x=y curve at three places. The mapping in Figure 4(c) was constructed the same as the one in Figure 4(b). Only three "points" are seen on the x=y curve. In going from Figure 4(b) to Figure 4(c), the system made a transition from chaos to a period-three state (the 1^2 state) through a tangent bifurcation.

A range of oscillatory behavior has been presented. There are many other types of oscillations possible including quasiperiodic oscillations. One might imagine that research would be undertaken to relate electrochemical oscillations to the oxidation process. In this sense we previously related the occurrence of oscillation under current control conditions of small organic molecules to a skeleton mechanism. There are at least two reaction paths in the oxidation of these molecules. One path, the direct route, corresponds to a relatively fast oxidation at low potentials. The other reaction path includes an adsorbed intermediate, surface CO, which reacts slowly at low potentials. It is only at a sufficiently high potential that surface CO reacts and is removed from the electrode surface.

Oscillations in current-control experiments can be explained as follows: Beginning at low potentials the reactions of the direct route initially satisfy the current. However, as surface CO builds up, there are fewer sites available on the electrode surface for the direct route. Consequently, the potential must rise so that the rate of reactions in the direct route increases so that the applied current is satisfied. The build-up of CO continues, as must the potential increase until a value of the potential is reached where surface CO begins to react and be removed at a rapid rate. With the removal of CO, the direct route has a large number of sites. Hence the potential decreases a smaller value is required to satisfy the current with the increase in the number of fast reactions.

Since the time we presented this formulation, others have repeated it. This formulation may be misleading. Oscillations occur in potential control experiments, but with a large solution resistance or a large applied cell resistance. Then, as in the current control experiments, there is an equation for the potential. Oscillations only occur when there is an equation for the potential. If the system is potential controlled and possesses a very small solution resistance, then there are no oscillations. These observations seem to imply that nonlinearities in the potential, and probably not only in the form of the Butler-Volmer rate coefficients, are responsible for the oscillations. Nonlinearities in the species, as might be implied by the above description are unlikely to be the fundamental cause.

4. Some Effects of Anions on the Electrochemical Oxidation of Formaldehyde

Anions affect the oxidation of any compound if the process has surface intermediates. [29,30] Anions inhibit reactions. In Figure 5 are 3 measured cyclic voltammograms for the oxidation of a 0.1 M formaldehyde solution (formalin) at a Pt electrode. Each voltammogram is for a 0.50 M single electrolyte. Perchloric acid was the electrolyte for the measured voltammogram with the highest current peaks and sulfuric acid was the electrolyte for the measured voltammogram with the second highest current peaks. Nitric acid was the electrolyte for the voltammogram with the lowest current peaks.

It is clear that nitric acid inhibits the electrochemical oxidation of formaldehyde more than does perchloric acid or sulfuric acid. The results explain why perchloric acid is the electrolyte of choice because it inhibits less than most other inorganic acids. The reason for this is that the perchlorate anion does not specifically adsorb. It maintains its hydration shell during the adsorption process and does not bond with the atoms of the electrode surface. Nitrate anions are adsorbed specifically and bond to electrode surface atoms.

Despite the typical results on anion inhibition, the results in Figure 6 show that the nitrate anion can enhance reaction. The solid curve is the voltammogram for a 0.10 M formaldehyde solution containing only 0.5 M perchloric acid as the electrolyte.

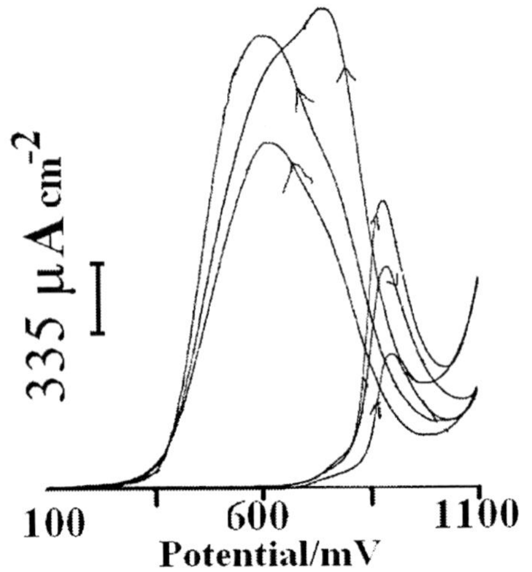

Figure 5. Current is plotted against potential. The curves represent three voltammograms for the electrochemical oxidation of formaldehyde in three different electrolytes: 0.5 M perchloric acid the highest peaks, 0.5 M sulfuric acid the second highest peaks, 0.50 M nitric acid.

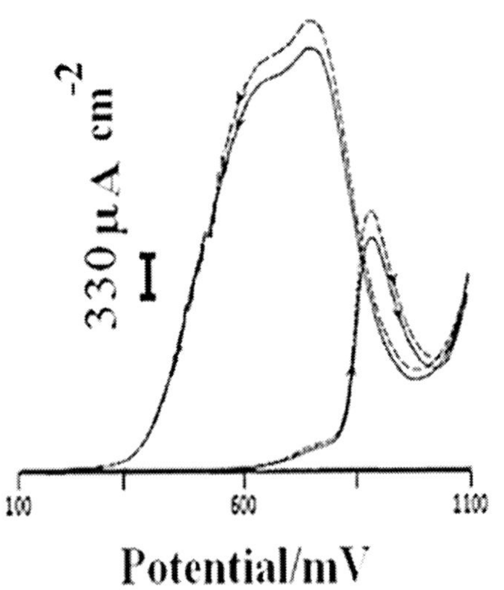

Figure 6. The solid voltammogram was measured using a solution containing 0.1 M formaldehyde and 0.5 M perchloric acid. The dashed curve is the same solution except .0005 M nitric acid was added.

The dashed curve is the voltammogram for the case in which 0.0005 M of nitric acid was added to the previous stated solution (the perchloric acid concentration is 0.495 M.) Somehow the current was enhanced on the addition of a small amount of nitric acid. The amount of enhancement in the current is plotted against the added concentration of nitric acid in Figure 7. The original theoretical idea for the enhanced current is as follows: Perchlorate anions are in excess. They will have a large surface density on the electrode. When a nitrate anion is adsorbed it bonds to the electrode surface. If the concentration of perchlorate anions is large enough, several of them will be in the vicinity of the adsorbed nitrate anion.

Ion-ion interactions are possible between the nitrate anion and the perchlorate anions. The nitrate is bonded so the interactions do not move it. However, the perchlorate anions are repulsed and leave the electrode surface causing sites to become available for the reaction of formaldehyde.

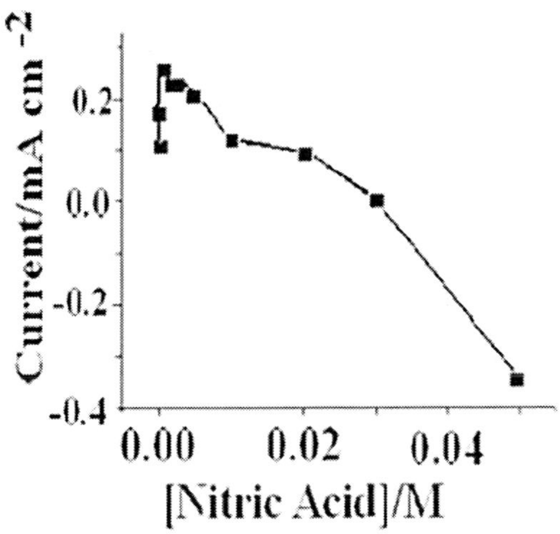

Figure 7. Plot of the increase in peak current of the forward peak achieved by adding nitric acid to a solution originally containing 0.1 M formaldehyde and 0.5 M perchloric acid.

This enhancement only occurs for small concentrations of nitrate anions, as their inhibiting characteristics will eventually overcome the enhancing properties as the concentration increases.

We found a problem with this explanation. We did the experiment in two ways. We increased the nitrate anion concentration by first adding to the solution represented by the solid curve in Figure 6 from a mixture containing 0.1 formaldehyde and 0.05M nitric acid. After reaching a concentration of 0.10 mM nitric acid we added larger increments from a mixture containing 0.25 M nitric acid, 0.1M formaldehyde and 0.25 M perchloric acid. The results in Figure 7 were obtained. However, redoing the experiments but leaving out perchloric acid in the second mixture causes a different result. Instead of the current enhancement decreasing past zero as in Figure 7, a substantial enhancement, although less than the maximum, continues past 50 mM nitric acid. This result is inconsistent with the explanation using surface anion interactions for two reasons. The additions with perchloric acid should give the same results, as those obtained with additions that did not contain perchloric acid except a larger amount would have been required. The second reason is 50 mM is a large concentration. The inhibiting effects of the nitrate anion should set in at this concentration. A solution of 0.1 M formaldehyde, 0.25 M nitric acid and a 0.25 M perchloric acid exhibits a current less than the current in Figure 1 for the solution containing only perchloric acid as the electrolyte. We hypothesize that another process occurs that is affected by nitrate anions. It is sensitive to the concentration of the anions and/or pH. Formaldehyde separates from methanol at the electrode surface where polymerization is possible. If nitrate is a less effective as a catalyst for polymerization compared to other anions, than a decrease in polymers will enhance the current. Otherwise, we have no explanation for the result.

CONCLUSION

Formaldehyde is well known as an important component in syntheses. We have given an overview of formaldehyde in electrochemistry. We have surveyed some electrochemical research on what might be now considered standard topics like sensors, examining changes in the electrochemical oxidation of formaldehyde by manipulating electrode surfaces and probing for surface intermediates with spectroscopy. Less studied phenomena was presented that included a variety of nonlinear behaviors exhibited by the variation of the potential as a function of time under galvanostatic conditions. Anion inhibition of the electrochemical oxidation of formaldehyde was discussed and an unexplained unusual anion effect was depicted.

REFERENCES

[1] W. Vielstich, Fuel Cells, Wiley, New York, 1970.
[2] R. Parsons, T. VanderNoot, *J. Electroanal. Chem.*, 257 (1988) 9-45.
[3] M. Islam, R. Basnayake, C. Korzeniewski. *J. Electroanal. Chem.* 599 (2007) 31-40.
[4] V.J. Cogliano, Y. Grosse, R.A. Baan, K. Straif, M.B. Secretan, F. El Ghissassi, *Environ. Health Perspect.*, 113 (2005) 1205-1208.
[5] L. Zhang, C. Steinmaus, D. A. Eastmond, X. K. Xin, M. T. Smith, *Mutation Research/Rev Mutation Research*, 681, (2009) 150-168.
[6] M. Hämmerle, E. A.H Hall, N Cade, D Hodgins, *Biosensors Bioelectron.*, 11, (1996) 239-246.
[7] S. Achmann, M. Hämmerle, R. *Moos Sensors* 2008, 8, 1351-1365.
[8] M. B. Ali, M. Gonchar, G. Gayda, S. Paryzhak, M.A. Maaref, N. Jaffrezic-Renault, Y. Korpan, *Biosensors Bioelectron.*, 22, (2007) 2790-2795.
[9] Z.-L. Zhou, T.-F. Kang, Y. Zhang, S.-Y. Cheng *Microchimica Acta*, 164 (2008) 133-138.
[10] Q. Yi, F. Niu, W. Yu, *Thin Solid Films*, 519 (2011) 3155-3161.
[11] K. Nishimura, K. Machida, M. Enyo, *J. Electroanal. Chem.*, 251 (1988) 103-116.
[12] J. Stelmach, R. Holze, M. Bełtowska-Brzezinska, *J. Electroanal. Chem.*, 377 (1994) 241.
[13] R.R. Adžić, M. L. Avramov-ivić, *J Electroanal. Chem.* 134 (1982) 177-180.
[14] S. M. Motoo, M. Shibata, J Electroanal. Chem., 139, (1982) 119-130.
[15] V. Selvaraja, A.N. Graceb, M. Alagarc, *J. of Colloid Interface Sci.*, 333 (2009) 254–262.
[16] S. Park,Y. Xie, M. J. *Weaver Langmuir* 2002, 18, 5792-5798.
[17] J. Geng, Y. Bi, G. Lu, *Electrochem. Comm.* 11 (2009) 1255-1258.
[18] T. *Solomun Surface Science* 176 (1986) 593-602.
[19] P. Olivi, L.O.S. Bulhões, J.-M. Léger, F. Hahn, B. Beden, C. Lamy J *Electroanal. Chem.*, 370 (1994) 241-249.
[20] P. Olivi, L.O.S. Bulhões, J.-M. Léger, F. Hahn, B. Beden, C. *Lamy Electrochim. Acta* 41 (1996) 927-932.
[21] Atsaushi Miki, Shen Ye, Takahiro Senzaki , M. Osawa, *J. Electroanal. Chem* 563, 2004, 23-31.
[22] M. Kikuchi, Y. Mukouyama, H. Okamoto, *Electrochim. Acta*, 53 (2008) 7817-7824.
[23] G. Zhao, Y. Tang, R. Chen, R. Geng, D. Li, *Electrochim. Acta,* 53, (2008) 5186-5194.

[24] M. Kikuchi, W. Kon, S. Miyahara, Y. Mukouyama, H. Okamoto, *Electrochim. Acta*, 53 (2007) 846-852.
[25] *Y-H Xu, M. Schell, J. Phys. Chem.*, 94 (1990) 7137–7143.
[26] M. Schell, F. N. Albahadily, J. Safar, Y-H Xu, *J. Phys. Chem.*, 93 (1989) 4806– 4810.
[27] H.F. Hunger, *J. Electrochem. Soc.,* 115, (1968) 492-497.
[28] M.T.M. Koper, M. Hachkar, B. Beden, *J. Chem. Soc.-Faraday Trans.*, 92 (1996) 3975-3982.
[29] J. Sobkowski, K. Franaszczuk, K. Dobrowolska, *J. Electroanal. Chem.*, 330 (1992) 529-540.
[30] D. V. Tripkovic , D. Strmcnik , D. van der Vliet , V. Stamenkovic, N. M. Markovic, *Faraday Discuss.*, 140 (2009) 25-40.

INDEX

C

D

E

I

J

K

L

M

N

O

P

Q

R

S

T

Q

R

S

T